20p

Parents start here

This chapter introduces parents to the prospect of their children having school homework to complete and offers some suggestions about the working environment.

Covers

Chapter One

Why homework?

For many 11 year olds, arrival at secondary school means, amongst other things, the introduction of homework. Although many will have had a limited amount of homework in their primary school, Year 7 pupils (11/12 years old) can expect homework on two or more subjects per night.

Most people are only at school once and so it makes sense to get as much out of it as possible.

Some people argue that homework should be abolished and an hour added onto the school day. That may happen one day, but at present homework really needs to be done to provide the best possible foundation for examination work.

Many children see homework as a chore that teachers inflict on them for no particular reason. As a consequence, homework is frequently not done properly, if it's done at all. The fact is, it would be virtually impossible to pass GCSE examinations without homework as there is simply too much material that has to be learnt in the time available.

Although this book is aimed at Key Stages 2 & 3 (i.e. 9–14 years of age), public examinations will be on the minds of pupils, teachers and parents at this stage, and the first three years of secondary school can be regarded as a foundation for examination studies in years 10 and 11. But before reaching GCSE courses, there are the SATs (Statutory Attainment Targets) at the end of the Key Stages. These tests also require preparation and once again homework will need to be completed if the highest grades are to be attained.

Your child(ren) will need to be provided with an environment in which serious study can be achieved. In order to work efficiently at home they will need two things:

Sitting in front of the TV whilst doing homework is not usually recommended if you want anything better than mediocre grades.

- The correct attitude themselves

- The correct attitude from their parents

If pupils see homework as being a chore, then it will be and as we all know, if your heart isn't in your work, the quality is likely to be below the standard it should be, or could be. The attitude of the child frequently comes from the attitude of their parents. That means parents need to provide support, encouragement and interest.

Support

Support means providing children with the wherewithal to complete their homework. Before considering splashing out on a computer, begin by providing a suitable place in which work can be done. And that doesn't mean an armchair in front of the television. There should be a place where a child can go where there are no distractions and where there is a table at which they can sit. One thing that's certain to create a chore from homework is having to 'set up' their homework facilities every night and then pack it all away when they have finished. Ideally, everything should be at hand. Including the computer.

Encouragement

This is not simply asking "Have you got any homework tonight?". It means helping to provide a schedule so that work can be done, yet still leaving enough time for relaxation. It also means providing the materials required (within reason) to complete the work.

Take an active interest in your child's homework.

Some schools will provide a homework timetable and it's worth spending time looking at it. It will have been planned carefully so that pupils are not given one piece of homework for each of three nights and then overburdened with it for the other two nights. There should be an even spread with most subjects being covered. You should also have an idea about the expected duration of homework. Clearly a piece of homework that takes one child 30 minutes to complete might take another a lot less. But you should have some idea of the overall time your child should be spending on homework each night.

Interest

This is always a difficult one where teenagers are concerned as they don't generally appreciate an interrogation about their schooling. But you should check to see that they are doing their homework and, if you suspect they are not because of the nightly cry "I haven't got any homework to do today", you should contact the school to find out if homework is being set.

Interest doesn't mean giving a cursory glance to a couple of pieces of work. Read what your child has written: you might even enjoy it!

What does s/he need?

What your child may *need* and what s/he may *want* are notions that are frequently separated by several hundred pounds. What you certainly don't need is a top-of-the-range computer capable of sending a space rocket to Mars. As consumers, we tend to buy computers in the same way that we used to buy hi-fi – we choose the models which have the largest numbers attached to them. A 100 watt music amplifier has got to be twice as good as a 50 watt amplifier hasn't it? With that misguided theory we buy a hugely powerful music system with no consideration of how the roof is going to stay on.

Computers have taken over the role of attracting customers by technical data and as a result half the country is operating a computer system with more power than they're ever likely to need. That said, ever more complex software places more and more demands on computer processing and memory and so you will need something a little higher than the absolute basic system. Most important is to buy a system that can be upgraded. In simple terms, the power or speed of a computer is measured by:

Buy a computer that is upgradeable so that when more power is needed it can be added.

Processor speed

This is the brain of the computer and the faster it is, the faster the computations.

Memory

Memory can be added to a system relatively easily and cheaply, so as long as you have the necessary amount to run the applications you want to run now, then that's probably enough.

Hard disk

The hard disk is the device onto which computer applications and work is stored. All computers are now fitted with a hard disk, but I favour adding a second at the outset – one for the applications and one for the work.

Two hard disks are better than one. If the data hard disk fills up, it is relatively easy to move your work onto a larger one, but not so easy to upgrade the disk with the applications.

Each of these has a direct bearing on the performance and cost of a computer system. Rather than buying the very best, it actually makes more sense to buy a system that is upgradeable so that, if and when the time comes that you need more power, you can add it to your computer.

Hardware

Which computer should I buy?

I've lost count how many times I've been asked this question over the years. The answer is simple: the best computer is the one that does what you want it to do. In other words, decide what you want the computer for and then choose the model that satisfies those needs.

Although the price of computer systems has dropped dramatically over the years, it's probably still one of the most expensive pieces of apparatus you'll find in most homes. It therefore makes sense to get the maximum amount of use from it.

That means that more than one person in the family might want to use it and it would be worth considering each individual's needs. Perhaps one or both parents use a computer at work and a home computer would enable them to do some of their work away from the office. But if the main purpose of the computer is to enable your child to complete homework then you should find out what type of computer is used at school.

Look at the type of computer used at school and consider buying a similar system.

There are two main types of computer system used in schools today:

- PCs running Microsoft Windows

- Apple Macintoshes

Of the two, the PC is the most popular in schools, as it is generally. For that reason, this book will deal mainly with a PC. Screenshots that appear throughout the book are from a PC which is running Windows XP (the latest incarnation of the operating system created by Microsoft) although all of the material (apart from specific technical details such as how to copy and paste a picture) could be applied to Apple computers.

The advantage of using the same computer at home and school is that material created at home can be transported into school to be worked on there. Although it is possible to convert files created on a Macintosh into a format that would work on a PC, it's not always as straightforward as some would have you believe and should be avoided if possible.

Software

Programs are referred to as software or applications. When buying a new computer, people tend to forget that software is needed. Just like a music system needs something to play on it, a computer needs software to work.

Some computers are supplied with 'bundled' software – programs the computer supplier pre-installs on new computers. Although it's free, you often find it's quite adequate for most school/home use.

If your computer isn't supplied with bundled software, you need to budget for it when buying the computer.

If you have a choice of software, once again, look at what the school is using and, if possible, buy the same. The reason is that all software generates files (the work) which are often in a special format readable only by that program and no other. A piece of work created in a program at home may not necessarily be able to be opened by the equivalent program in use at school, just because the program you are using happens to be by a different manufacturer than the one being used at school.

If your computer was supplied with software, check what you have. If not, buy the same as the school.

Many programs can save files in a variety of formats. For example, if your school is using Microsoft Word (a word processor) it would be my first choice to buy the same program to use at home. That would mean that work could be freely moved between home and school on a simple floppy disk. The advantages of this arrangement are obvious: work begun at school can be continued at home and vice versa. But if you use StarOffice at home you will be in a similar position because that program will save files which can be read by Microsoft Word. Equally, Microsoft Word documents can be read by StarOffice. This file compatibility works reasonably well in many cases, but it's not always the perfect solution. Although StarOffice can read Microsoft Word files, it sometimes fails to get the layout exactly right. In other words, compatibility is not always 100% assured.

Word processed files transport better than many. Some programs output files that simply cannot be read by any other program. In these cases the only solution is to get the program being used at school if you want your child to be able to move work between home and school.

Office suites

The so-called office suites of applications contain the most commonly used applications and are the ones mainly used in this book. They include a word processor, spreadsheet with graph drawing facility and a database.

Microsoft Office is probably the most well-known of the office suites. There are several versions ranging from Basic to Professional and many schools use it because there are special rates for education.

Look out for special deals for students.

Microsoft Works is a simpler version of Office (and consequently cheaper) but contains most of the key features of MS Office. This is sometimes seen bundled with new computers.

Computers are often supplied with Lotus SmartSuite pre-installed or with the CDs packaged with the documentation. This is a very capable tool and will provide you with all of the key features likely to be needed.

Also, consider buying Paint Shop Pro by JASC Software (www.jasc. com).

Paint Shop Pro has many of the features of high-end paint programs like Adobe Photoshop, but at a fraction of the price, and is much easier to use.

Another possibility is StarOffice which is actually free if you download it from the Internet. (Visit www.sun.com/staroffice/.) To be more precise, it will cost you what it costs for the phone call whilst you're online downloading it. If you don't fancy spending a couple of hours watching it being squirted down the telephone line, you can buy it very cheaply on CD ROM. It includes everything you're likely to need, including a drawing program, email support and a browser for you to access the Internet.

Painting and drawing

Microsoft Windows is supplied with a painting application called Paint. Paint is adequate for most purposes but not if you want to modify pictures taken with a digital camera. For this a special application is needed and one of the best budget titles is by Greenstreet. You can find out more about their photo editor by visiting the GST website at www.gstsoft.com/.

Whilst visiting the GST site, you might also look at their drawing program (called Draw) which is another popular program which will be used for drawing (as opposed to painting).

The Internet

It's this feature of computer life that has generated so much debate in recent years, not only because of unsuitable material distributed via the Internet but simply the cost of accessing the Web. Most computers are supplied with Internet access built in. To access the Internet you need an Internet account, a modem, a phone line and a browser.

Don't become too preoccupied with the negative side of the Internet. There is a lot of good stuff out there as well.

Internet account

This is effectively your licence to access the Internet and takes the form of an agreement with an Internet Service Provider (ISP). There are many ISPs, some from very unlikely sources, and there are many different payment plans available. For example, many provide a free-call number for Internet access and offer unlimited unmetered access to the Internet for a low fixed monthly fee.

Modem

This device may go into the computer or sit on the desk as a separate item. Either way, it connects the computer to a telephone line. If you haven't got one in your computer, they are very cheap and a typical modem will have a speed rating of 56k/second.

The phone line

For the desktop computer with a modem, you'll need a BT-type phone socket near to the computer or a long phone extension lead to connect the modem to the phone line. Kits are available that let you add a socket to an existing phone installation without having to open the phone sockets.

It's not really practical to attempt to carry out serious research if you're constantly worried about running up a large phone bill, so look for an ISP that offers special connection plans to tie in with when your children are most likely to need Internet access. But remember, whatever deal you have, whilst you're online (i.e. connected to the Internet), the phone will be in use. This means that no one can make a call from the phone, and anyone calling you will find the line is engaged.

A second phone line is worth considering if you intend spending long periods online.

The cheapest solution to part of this problem is to ask your phone company to provide you with a voicemail service which enables callers to leave messages if your line is engaged. But to really overcome the problem (if indeed it is a problem) a second phone line is required. Many phone companies offer very attractive deals for second lines.

If you want to free-up the phone *and* provide a better Internet service, consider signing up to a Broadband service. (Visit www.blueyonder.co.uk/ for more details about Broadband.)

Browser

The fourth requirement for Internet access is a browser. The browser is the program that essentially does two things. Firstly, it is the software which enables the user to commute to different websites throughout the world. Secondly, it displays the pages of information from the websites.

In simple terms, the browser decodes documents written in a special language called HTML (Hypertext Markup Language). Of all the browsers produced since the Internet first became widely used, two have established themselves as supreme: Internet Explorer (which is supplied with Windows) and Netscape Navigator (now tied up in the total Internet package called Communicator).

Under attack?

Regrettably, there are certain individuals whose sole aim appears to be to destroy. These are the people who develop and distribute viruses, which in many countries is a criminal offence.

The more you use the Internet, the more chance you'll have of collecting a virus. Some are fairly harmless and little more than a minor annoyance. Others are vicious and spiteful and can do a great deal of damage to your computer software, including (but not limited to) corrupting or erasing all your work, stealing passwords and other account details and using your email addresses to send incriminating/defamatory/offensive material from your email address to your friends and colleagues.

Software to protect you from virus infection (and variants like worms) is a minimum requirement especially if disks containing work are to be moved between home/school. If you have a permanent connection (like Broadband), a firewall to prevent outsiders accessing your work is also essential. Visit Symantec at www.symantec.com/ for details of protection solutions.

It's wise to monitor your child's activity on the Internet.

Clearly, viruses which will destroy work, including school work, must be detected before they can do any damage.

Use a virus protection program that can automatically download the latest virus definitions so that you are always fully protected against the latest viruses.

Using a computer

This book suggests ways of completing homework using a computer. However, it should be noted that not all homework *can* be done using a computer and not all homework *should* be done using a computer. But a great deal of it can, and I observe from my children that the use of a computer makes the completion of homework more interesting and consequently less of a chore. As a result, my children's homework grades are consistently higher than the grades they attain in their classroom.

Although the title of this book refers to homework (usually, a piece of work done in a single evening), many subjects require coursework which will take the form of an extended study over several weeks/months. The ideas in this book will be equally valuable for 'traditional' homework and more detailed assignments.

This book is divided into two sections:

Software (chapters 2 to 7)

The first section provides general help with the programs most likely to be used to complete homework. It gives an overview of the main programs, what each does and which subjects are most likely to use them. It goes on to show how to research a topic and how to include research material in a piece of homework. There then follow some tips on how to lay out work ready for printing and finally how the work is best printed, either at home or school.

Curriculum issues (chapters 8 to 20)

The second section covers the main curriculum subjects where homework might be given and provides typical examples of the sort of homework that a child might be given which could be completed using a computer. It is, of course, impossible to predict every type of homework that could be given for each subject, but it gives examples of the sort of computer-based work that is likely to be needed. At the end of each subject chapter is a section entitled *No homework?* which outlines a simple subject-related task which will provide some additional IT skills as well as developing some material which could be valuable to the subject being studied.

Using a computer to complete homework and assignments should make the whole notion of completing homework much more palatable and therefore less of a chore.

General help

Using the right tool for the job is vital if you are to use your computer effectively. Learn what each program does and where it is best used.

Covers

Chapter Two

Word processing

Without doubt, the word processor is the program you will use the most. This applies not just to school work, but probably throughout your working life.

A word processor is the computer equivalent of a typewriter. It even uses the same input device – the keyboard. The reason the keys on the modern computer keyboard are in the order they are dates back to early mechanical typewriters. They were laid out in what we now call QWERTY pattern because putting the keys in alphabetical order meant that some letters which were next to each other frequently caused the typewriter mechanism to jam.

There is another difference. With a typewriter you have to change lines when you get to the end of the page. With a word processor you don't.

There is one major difference between a word processor and a typewriter and that is that the typewriter commits the text to paper immediately. The word processor stores it in the computer's memory first and then puts it onto paper at the very end.

This means that once you've typed a document on a typewriter, that is it. You can't make any changes to it other than using correction fluid to blot out one or two words before retyping the corrected word. Any more than a couple of corrections and the page begins to look quite horrible. You certainly cannot move chunks of text around the page as you can with a word processor.

You will use a word processor for most assignments that require text. That probably means that most of your homework will be done using a word processor, so it makes sense to learn how to use it properly.

If you have Windows then you have quite a good word processor called WordPad. It doesn't have all the features of the very latest word processors like Microsoft Word, but it can tackle most of the things you're likely to need.

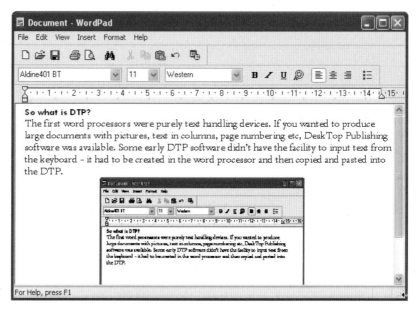

Most DTP work can be done using a good word processor.

So what is DTP?

The first word processors were purely text handling devices. If you wanted to produce large documents with pictures, text in columns and wrapped around a picture, page numbering etc., you had to use a DeskTop Publishing program as word processors couldn't do that type of work. Early DTP software was also fairly basic. Some didn't even have the facility to input text from the keyboard – it had to be created in the word processor and then copied and pasted into the DTP.

Things have moved on since then and word processors can do so much more. In fact, a good word processor like Microsoft Word can do all the DTP work you're likely to need to do at school. Even WordPad can handle graphics, use different text styles and have objects like tables embedded in it, so you can do quite a lot of DTP-type tasks even without a top-of-the-range word processor or DTP program.

It's important that you learn how to use a word processor. You need to learn how to copy blocks of text from one place to another. Different programs work in slightly different ways, but in most cases, copying and pasting is carried out in the following way. Suppose you want to move paragraph 1 so that it is between paragraphs 2 and 3:

1 Move the mouse pointer to the beginning of paragraph 1, click and hold the left button and whilst holding it move to the beginning of paragraph 2. Release the button. The text will invert and we say that that section is highlighted.

If you want to make a copy of a section (keeping the original in place) use Ctrl+C instead of Ctrl+X.

2 Press Ctrl+X which will remove the marked section from the document and will place it in a special area of memory called the Clipboard.

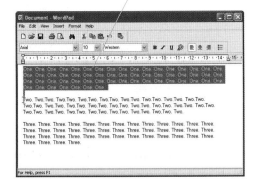

3 Move the mouse pointer to the beginning of paragraph 3, click the left mouse button and press Ctrl+V to paste in paragraph 1.

Some word processors allow you to 'drag and drop'. After marking a section, drag it to where you want it to go.

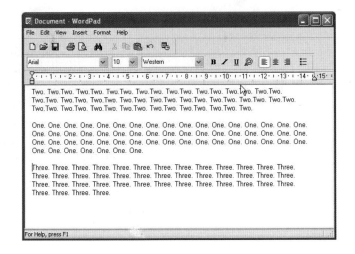

Spreadsheet

A spreadsheet is a very clever program that is designed for handling numbers. Originally it was designed specifically for money and so accountants used it.

You will use a spreadsheet whenever you want to create a table of figures and/or text. Maths and science would seem to be the most likely subjects, but because you can use words as well as numbers, it could also have uses in other subjects like technology where you might want to create a list of parts.

When you open a spreadsheet you are presented with a page full of boxes called cells:

 If you haven't got a spreadsheet, you can download StarOffice free from `www.sun.com/staroffice/`.

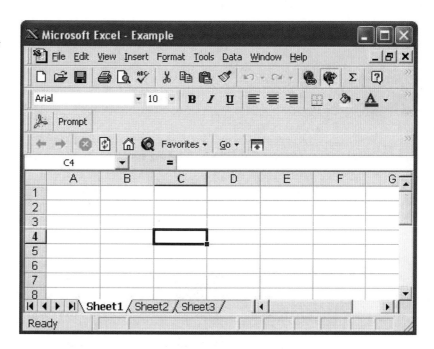

You'll notice that across the top of the cells are letters beginning with A and down the side are numbers beginning with 1. These enable you to identify each cell. Clicking on a cell highlights it and the cell currently highlighted (above) is C4 (not 4C).

Into each cell you may enter a number, word, formula or graphic.

Some spreadsheets have slightly different ways of entering a formula, but this works with most popular programs.

 Changing one or more pieces of data to see what happens is called 'What If?'. For example, What If we change the 5 to a 4?

 Once you've entered a formula, you don't have to enter it again and again for every cell you want to use it in. Select the cell with the formula, press Ctrl+C, click and drag over the cells you want to copy to and press Ctrl+V.

Here's a quick exercise to help you understand what a spreadsheet can do:

1 Move the mouse pointer to cell A1, click the left button, type *5* and press Enter.

2 Click in cell B1 and enter *7*.

3 Now go to cell C1 and type =A1*B1 and press Return.

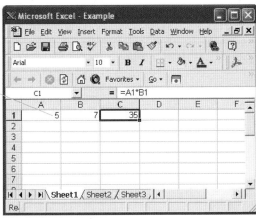

You'll notice that what appears in cell C1 is not what you entered, but the number 35. What you've done is to tell the software to multiply A1 and B1 and put the answer in C1. Nothing very clever there, you might think. But if you change the number in A1 to 4, when you press Return the number in C1 will also change. This is the clever part of a spreadsheet, it can automatically update itself and do lots of calculations based on just a few numbers.

Creating a graph

The other feature of a spreadsheet is that the software usually lets you turn your data into a graph. Generally, depress the left mouse button whilst you drag the pointer over the data you want to graph.

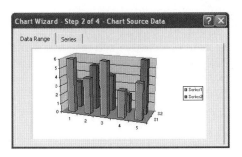

Select the Graph button and choose a graph type. These are usually Line (for a single, changing value like temperature), Bar (for comparing several quantities) and Pie (for showing fractions or parts).

Database

If you haven't got a database, you can download StarOffice free from `www.sun.com/staroffice/`.

There are many good database applications. A popular one is DataPower by Iota (`www.iota.co.uk/`). A database is a collection of organised information. You've probably got several databases at home. The telephone directory, for example, is a database. The problem with a database in book form is that you're limited to what information you can get out of it. The information in a telephone directory is ordered by people's surnames in alphabetical order. If you want to find D Brown's phone number, you open the directory at 'B', then look for 'Br' until you find all the people whose name is Brown. You then search down the first names until you get to 'D'. When you've found the correct name, you can copy down the phone number.

The precise method for creating a database will depend on which database program you are using.

If you have a phone number in your pocket, the telephone directory will be of no use in finding the name of the person whose number it is, unless you're prepared to read every entry in the directory, starting at 'A', until you find the number. Clearly not a practical proposition. But with a computerised database you could do exactly that.

The information in a database is held in categories called 'fields'. If you wanted to create a database of your friends' contact details you might have fields like surname, first name, address, home phone number and mobile phone number.

The final layout might look something like this:

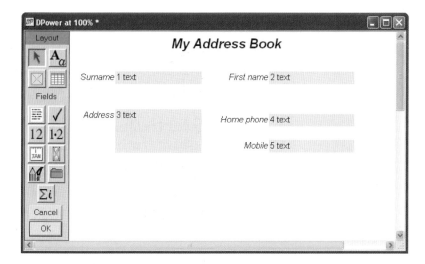

Each field may contain either text, (a combination of letters and numbers), a whole number (called an integer) or a real number (which may have decimal places). Different database programs may also feature date fields, time fields and pictures.

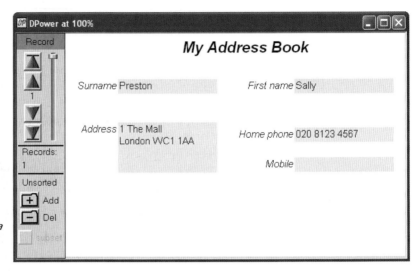

A database is featured in many organisers. If you have such a device, an address book like this is really useful.

Once you've created the layout, you can then begin entering details in the fields. In time you will have built a large database of friends, family and acquaintances.

Searching

Again, different database applications will handle searching in different ways. But whatever program you are using, you would be able to search for any piece of data. For example, if you asked it to search for 'Sally' it would display the card (or record, to be technically precise) shown above. If there were more than one person called Sally it would show that record too.

If you included a field called 'Birthday' you could search for all people who have a birthday on a particular date.

Sort

You can also sort records into any order you like. The most likely order in the case of an address book would be by surname. But you could sort them by phone number if you wished.

Graphics

There are two distinct types of computer graphic:

- Bitmapped

- Vector

The computer screen is made up of tiny dots called *pixels* (short for picture elements). Bitmapped pictures simply apply colour to each of the pixels to produce a picture. To alter the picture, you simply change the colour of some of the pixels either by hand or by using a program with tools to re-colour certain areas.

Although you can display a digital photo in a basic painting program, there's not much you can do with it. Try and find a good photo editor if you want to work with pictures like this.

Bitmapped pictures are produced using 'painting' programs like Microsoft Paint (which is supplied with Windows), by scanning a picture with a scanner or by using a digital camera (above).

The advantage is that they can hold lots of detail, but there are several disadvantages. First, the files can be very large indeed – it's not difficult to create a picture that uses so much memory it is unable to fit onto a floppy disc. Second, you can't easily change the composition of a bitmapped image. Just like a photograph taken with a film camera, you can't decide that you want to remove part of the composition after the picture has been taken as the initial picture contains no information about what is behind objects or people.

Third, bitmapped images don't scale very well. If you try to scale down a picture taken with a digital camera you will lose some of the detail. If you try enlarging it, all that happens is that the pixels appear to get larger and you finish up with chunky graphics. (A curve or a diagonal line will begin to look like a staircase.) Consider this simple bitmapped picture which has been doubled in size:

As you can see, it's not very smooth.

All versions of Windows come with this simple bitmap editor. Although you can use it to display pictures taken with a digital camera, you can't do very much to the image other than crop it, but you can use Paint for simple diagrams. If you want to edit digital pictures you need a photo editing program.

Vector Graphics

Vector graphics are completely different. A vector graphics program, sometimes called object-oriented graphics, is used for drawing rather than painting.

The principle is simple: a drawing is composed from a collection of simple components called objects. Each object can be moved independently of other objects, including placing objects behind others.

Objects can be freely scaled without losing any of their detail. This is the same drawing, as shown opposite, scaled by the same amount, but this time it's displayed as a vector drawing:

Vector and bitmap are the computer equivalent of drawing and painting.

Clearly, a much smoother result.

Draw

There are several drawing programs which can be bought quite cheaply from good computer stores. One of most popular budget titles is Draw by GST. (Find out how to buy this program from www.gstsoft.com/.)

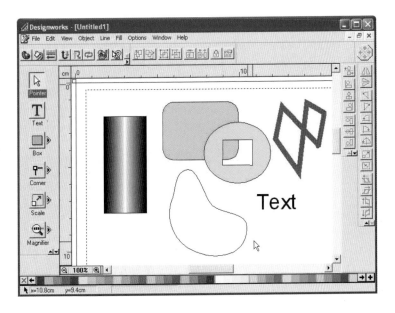

To recap...

These are guidelines, not definitive answers.

- If you want to produce lots of text, use a word processor.

- If you want to produce text with pictures and/or tables you need either a good word processor or a DTP application.

- If you want to perform lots of calculations, make changes to the original data and then recalculate, use a spreadsheet.

- If you want to compare sets of data in similar formats, you need a database, although a good spreadsheet can do much of the work a database does.

- If you want to search though data to find a particular item or items, you need a database.

- If you want to analyse data and ask 'What If?' questions then you need a spreadsheet.

- If you want to paint pictures rather like using a paintbrush on canvas, you need a bitmap painting program.

- If you want to play around with digital photos, you need a photo-retouching program.

- If you want to draw technical pictures and diagrams and be able to control each individual element, then you need a vector graphics program.

- If you lack one or more of these programs, download StarOffice by Sun Microsystems (`www.sun.com/staroffice/`) free of charge. The file is very large but provides just about everything you need including vector graphics, a word processor, spreadsheet and browser.

Organisation

A computer is a tool which will enable you to produce large quantities of work. Unless you carefully organise your work, you'll have great difficulty finding it later.

Covers

Chapter Three

Creating directory folders

You may be lucky and have a computer all to yourself but it is more likely you'll be sharing it with others in your family. Regardless of your personal situation, it is most important that you organise your work carefully. That way, you'll be able to find what you want quickly.

Central to organising computer files is the use of directory folders:

Spend time organising your work carefully before you begin.

A directory folder is very much like a real paper envelope folder into which you can place documents before storing it in a filing cabinet.

If you're one of several people using a computer in your house, you might begin by creating your own personal folder which could carry your own name:

Geoff

To create a new folder, first open My Documents (normally on the Desktop) by moving the mouse pointer onto it and double-clicking the left button.

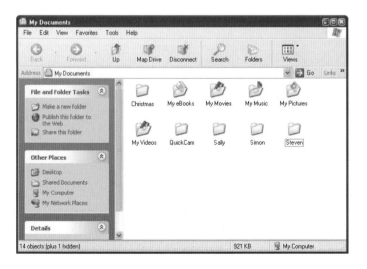

2 In a piece of empty space, click the right-hand mouse button which will open a menu.

3 Move the mouse pointer to New which will open another menu.

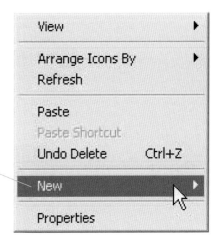

View	▶
Arrange Icons By	▶
Refresh	
Paste	
Paste Shortcut	
Undo Delete	Ctrl+Z
New	▶
Properties	

4 Choose Folder.

Folder
Shortcut

ART Image
Briefcase
Bitmap Image
Corel Photo House Image
Microsoft Word Document
DataPower Document
Design Tools File Type

X3D Document
CorelXARA Document
Microsoft Excel Worksheet
WinZip File

5 The new folder appears roughly where you first right-clicked:

New Folder

The name (New Folder) will already have been highlighted and so you may immediately type in a new name for the folder, after which you press Enter or Return.

Inside your new folder you may place further folders. These might include one called School Work. Inside that folder you could place a folder for every subject you study.

The route or path you have taken from My Documents is shown in the address bar.

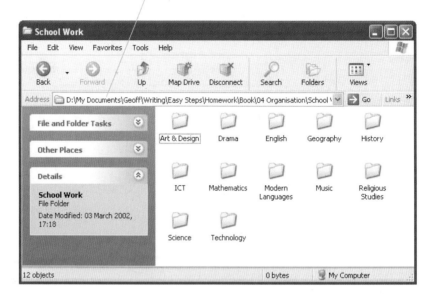

If you use Microsoft Windows XP you can customise these folders so that they each have a different icon.

You should be able to find a suitable icon for most subjects, but not all of them will include a 'folder' as part of the picture like the Music folder.

1   Right-click on a folder icon and choose Properties from the menu.

2   Choose the Customize tab at the top and then choose Change Icon from the dialog.

3   Choose the icon you wish to use and click OK.

Having created folders for your subjects, each piece of work you do should be placed into the appropriate folder. If you need to find, say, a piece of geography work you did a couple of months ago, you'll know exactly where to find it.

Naming files

The second important organisational lesson is learning how to name your files so that you don't have to open every piece of geography work in order to find the piece you want to refer to.

You don't need to include the subject title in the filename if you keep it in the correct subject folder.

You can find the date a document was produced by right-clicking on the file and choosing Properties from the menu, but having the date as part of the filename means that you immediately see when all of the files were produced as soon as you open the folder.

Clearly it is not very helpful if you're looking for a piece of geography work and you open the Geography folder to find this! You will have little or no idea what each piece is about. But if you used filenames like 'Erosion 181002' then you'll know that that file contains work on Erosion and it was done on 18th October 2002.

Sometimes a single piece of work may have several files. For example, you may have a piece of science work that includes a diagram and a table of results as well as a text document. In cases such as this, create a new folder within the appropriate subject folder and then save all the various files for a single piece of work inside.

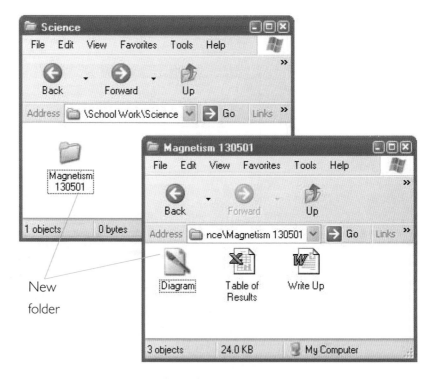

New
folder

Creating a new document

Creating a new document is exactly the same as creating a new folder, except you have to choose which type of document you want to create.

Open the appropriate folder (e.g. if you're beginning a piece of English work, open the English folder.)

2 In a piece of empty space, click the right-hand mouse button to open a menu.

3 Move the mouse pointer to New which will open another menu.

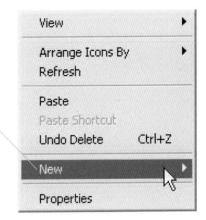

View	▶
Arrange Icons By	▶
Refresh	
Paste	
Paste Shortcut	
Undo Delete	Ctrl+Z
New	▶
Properties	

4 The programs installed on your computer will determine exactly what this menu will look like, but it will be similar to this.

5 Click on the type of document you want to create and it will be created in the folder with a default name (something like New Text Document).

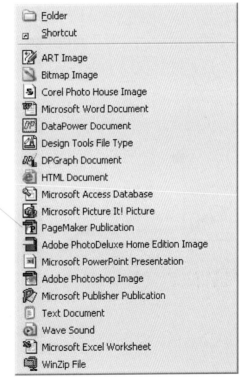

- Folder
- Shortcut
- ART Image
- Bitmap Image
- Corel Photo House Image
- Microsoft Word Document
- DataPower Document
- Design Tools File Type
- DPGraph Document
- HTML Document
- Microsoft Access Database
- Microsoft Picture It! Picture
- PageMaker Publication
- Adobe PhotoDeluxe Home Edition Image
- Microsoft PowerPoint Presentation
- Adobe Photoshop Image
- Microsoft Publisher Publication
- Text Document
- Wave Sound
- Microsoft Excel Worksheet
- WinZip File

6 Change the name in exactly the same way as you did when you created a folder and then double-click it to open it.

Saving your work

Saving your work regularly is vital. Mishaps like electricity failures and computer problems like hard disk failure are thankfully rare, but they do happen. If somebody accidentally switches off your computer, all unsaved work will be lost. If the last time you saved your work was 5 minutes ago, then it should take no more than 5 minutes to redo it. If the last time you saved your work was 5 o'clock and it's now 11 o'clock and you've been working solidly, it will make your eyes water!

Get into the habit of saving your work every 5 minutes or so.

I've lost count of the number of times I've seen work lost because it hadn't been saved. Yet the process is so simple.

Saving work is the same for virtually all Windows programs. If you created a document in the manner previously described, then pressing Ctrl+S will save it.

Duplicating files

Sometimes you may want to save a copy of your work in either a different location or with a different filename. In this case:

1 Click File at the top left of the program.

2 Click Save As...

Some programs have an AutoSave feature which can be set to automatically save your work every few minutes.

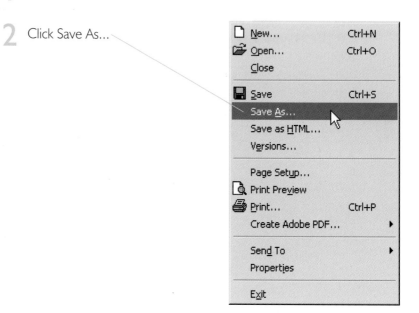

3 In the Save As... dialog, enter a new name and/or new location.

Research

You often need to find additional information yourself. There are several places where you might find what you're looking for, including the Internet and CD-ROMs, but sometimes you may need to contact someone.

Covers

Chapter Four

The Internet

If you have Internet access at home you have a very powerful resource that should help you find out about any subject you are asked to study for homework.

There are two ways to find out what you want to know:

- by typing in a Web address

- by performing a search

Begin by opening your browser (which in most cases will be Internet Explorer or Netscape Navigator) and establishing an Internet connection. The page that appears first will be your Home page and clicking on the Home icon on the button bar at the top of the window will always bring you back to this page.

Alongside the word Address is a panel into which you can type an address. Click on the panel to delete the current address, type in the address you want to visit and click on the Go button.

Dotted around the page will be words which are underlined and possibly in a different colour. These are called hyperlinks; clicking the left mouse button on one will take you to another page.

Sometimes, as you move the mouse pointer over the page, the pointer will change from an arrow to a hand. This also indicates a hyperlink. In the case of the Web page shown opposite, each of the book covers are hyperlinks and clicking on one of them will open another page giving additional information about the book.

Searching the Web

It's more likely that you won't have a specific address to type in and so you'll need to search for the information you want.

Suppose, for example, you want to find information about The Black Death as part of your History homework.

In the address bar in the browser, type in Black Death and click the Go button. The browser now displays a number of sites that feature the Black Death. It also lists pages about Death and about the colour Black, so you'll need to be selective:

Before starting a piece of homework, look on the Internet for information about the topic.

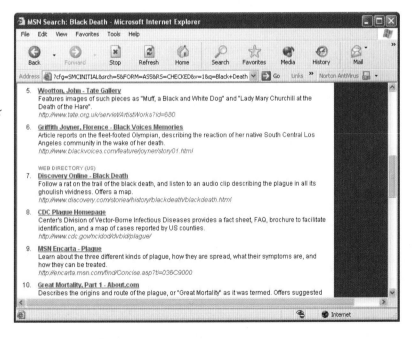

Read the paragraph alongside each heading and when you have found something that looks like it's going to provide you with the information you're looking for, click the title of the website shown (usually) in blue and underlined.

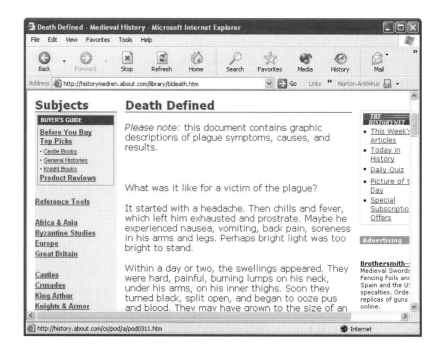

One of the options from the list of search results on page 39 linked up to the page shown above. When opened it was found to contain all the information needed about the Black Death. The page also contains links to other sites should you wish to find out more about it.

Information found on the Internet is not always accurate.

If the page you get doesn't provide you with the information you want, click the Back arrow (the arrow at the top left of the browser which points to the left) to take you back to the previous page. In this case, it takes you back to the list of search results where you can scan through and choose another website to visit.

Sometimes when you enter a search word you get thousands of results – too many to be able to work through. On these occasions you need to be a little more specific about your requirements. You can enter several words and so be fairly exact about your requirements.

Useful websites

There are several really useful websites which can provide very specific help for you when completing your homework.

BBC Bitesize

`www.bbc.co.uk/education/ks3bitesize/`

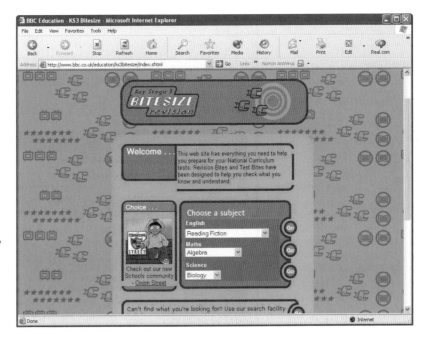

Put this site in your Favorites.

The Bitesize website is just part of the BBC's education provision which includes books and, of course, television programmes.

The site focuses on the Key Stage tests and so at present the material is for English, Maths and Science.

Although this site is for revision for the Key Stage tests, there are lesson notes which will provide information for you to use in your homework.

KidsClick

`www.kidsclick.org/`

This huge index and search site should be able to lead you to find those elusive answers to homework problems.

Channel 4

`www.channel4.com/learning/secondary.html`

The Channel 4 website provides material for all subjects and presents the material in an exciting way.

When you get into this site, choose the subject you want to research from the row of coloured tabs under the main heading.

Search engines

You can search for sites simply by entering a name in the address bar of your browser, but that's not the only way to get answers. There are many search engines which can be used to find pages; some will allow you to enter a specific question. Try Ask Jeeves at `www.ask.co.uk/`. You can type in questions like "What is H2SO4?" and Jeeves will return a list of sites which will provide the answer. Try entering "Where can I get homework help?"

CD-ROM encyclopedias

Another good way to collect information is from one of the many CD-ROM encyclopedias available from high street stores.

Even if you have Internet access, a good encyclopedia is worth having.

These products contain lots of information which, unlike the Internet, are likely to be better written and contain fewer mistakes. Most allow you to search alphabetically in exactly the same way as you would if you were looking up something in a book encyclopedia, but you can search for a key word or words which will display several articles.

Most CD-ROMs allow the users to copy and paste text and pictures into another document. It should be noted that this approach will not necessarily improve the understanding of the topic being studied.

The contents of a CD-ROM can't be changed so some information may be out of date.

The big difference between traditional book-based encyclopedias and the modern CD-ROM equivalents (like Microsoft Encarta shown here) are the 'extras' you get with the CD versions. Rather than still pictures, many articles feature moving pictures (animations/video) and sound. These additional features provide a much greater understanding of the topic being researched.

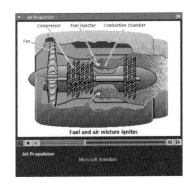

There are some encyclopedias which are subject-specific – they deal with one subject (e.g. Space, Periodic Tables, etc.) in detail rather than giving brief details about a wide range of subjects.

Capturing text and graphics

Having found the information you want, you will need to extract the information you need for inclusion in your work.

Text

In most cases, text can simply be marked, copied and pasted into another document.

> 1 Once you have found what you're looking for, highlight the text by moving the mouse pointer to the beginning of the passage, holding down the left mouse button and dragging to the end of the passage.

Ctrl+C and Ctrl+V are used by almost all programs to copy and paste.

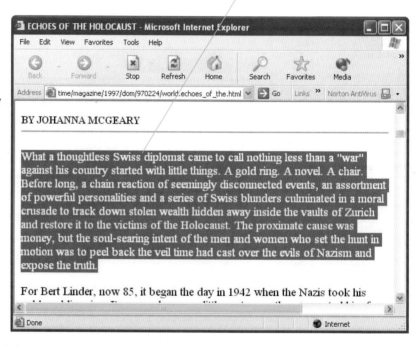

You will see that the text 'inverts' (the words change to the background colour and the background around the text changes to the colour the text was).

> 2 Next, copy the text by pressing Ctrl+C. This places the text in a special area of the computer's memory called the Clipboard.

3 Now open a new document using WordPad or another word processor. Move the mouse pointer to where you want the text to be placed, click the left mouse button once and press Ctrl+V to paste the text into your document.

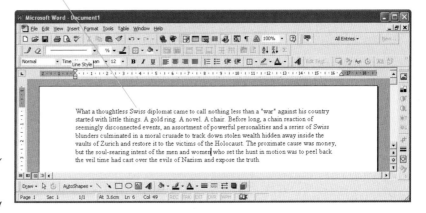

Copying text from the Internet or any source will not help you learn about the subject unless you read what you've copied.

This method has been explained to you so that you can disconnect from the Internet and study the text at your leisure without running up a large phone bill. But you should not attempt to submit text captured in this way as part of your homework. There are three reasons for this:

• It's not your work and it would be quite wrong to pass it off as such.

• Any teacher would immediately recognise that it's not your work because the writing style would be different.

• You will have learnt nothing about the topic on which you have been asked to complete the homework.

What you should do is read the text you have captured and rewrite it in your own words. That way you will not be passing someone else's work off as your own and you will have a better chance of learning what the topic is about.

Graphics

Capturing pictures or charts from the Web is equally straightforward:

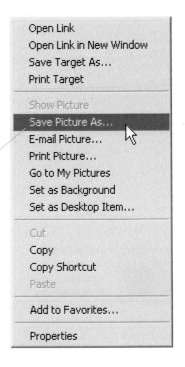

1 Move the mouse pointer onto the image you want to capture and click the right mouse button to open a menu.

2 Select Save Picture As...

3 A dialog will open giving you the option of choosing the filename and the location to save the picture.

Some versions of Windows will display a small graphic menu at the top of the picture when you move your mouse pointer over the picture. Clicking on one of the four icons will save the picture (the equivalent of right-clicking and choosing Save Picture As... from the menu), print the picture, email the picture or open the My Pictures folder.

Any pictures taken from the Internet should be credited in your work –
e.g. 'This picture was taken by [name of photographer] and was captured from [name of source]'.

Once the picture or diagram has been captured it can be included in your work. (See page 54)

Don't be afraid to list all sources of material used for the completion of a piece of work. A good list of sources will demonstrate that you have thoroughly researched a subject and not just looked at one source and collected all of your material from there.

Asking for information

Sometimes you might want to ask someone about a particular topic as you can't find anything about it on the Web. You often find a button labelled Contact Us on a Web page and this is your opportunity to ask a specific person a particular question.

When you write to someone from a 'Contact' link on a Web page, it's often best to prepare the text offline. In other words, use a text editor like WordPad to write your message.

1 Open a text editor and write your message. Check it carefully ensuring there are no mistakes.

When sending emails to individuals do NOT send your home address or telephone number.

2 Highlight the entire text (usually by pressing Ctrl+A) and copy it into the Clipboard using Ctrl+C.

3 Open your email program or open the Web page and paste the text into the message space using Ctrl+V.

Email

When sending an email to a business, you should make your message as brief as possible but you should remember to observe a few courtesies.

1 If possible, try to find out the name of the person you are sending the message to. This will usually be part of their email address e.g. gpreston@word4word.uk.com.

Even though email is fast, do not expect a reply the same day.

2 In this example, the person's surname is Preston so begin your message Dear Mr. Preston. Never write to a business using a first name unless you have been told that you may.

If you're sending a new message from your email program, stay offline until the message is complete and ready to send.

3 Begin a new line and write your message. Remember, *you* are asking *them* to do something for *you* which is unlikely to be of benefit to *them*. Be polite.

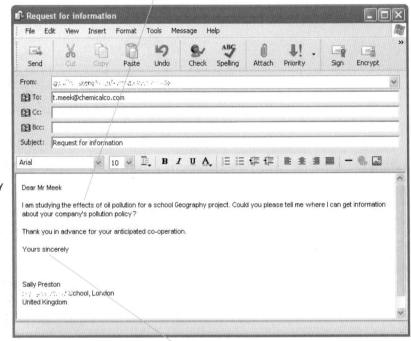

If you are asked for an address to send some printed material, give your school address, but check with your school first.

4 When finishing the message, use Yours sincerely (with a lower case 's' in sincerely), miss a couple of lines and type your name.

5 On the next line, type the name of your school and the town or city it is in. If you think the message may go overseas, type your country on the next line.

Don't expect a reply within minutes. You may get a quick reply but it's not always the case. If you have heard nothing for a week, you could try re-sending, but if you still get no response, give up.

Presentation

Once you've collected all the material for your homework you'll need to put it together in a way that can be given to your teacher for marking.

Covers

Chapter Five

Layout

Much of your homework will be completed using a word processor.

Modern word processors contain lots of features which will help you lay out your work in a way that will make it much more enjoyable to read than simply line after line of text.

Use the features in modern word processors with care.

Font styles

It is easy to change both the design and size of the characters used in your work. The design of the letters is known as the font and there are thousands of designs available. In most cases, clicking the arrow alongside the current font name will select that font. If you want to change the font style of a piece of text already entered, select the text and then choose the font style.

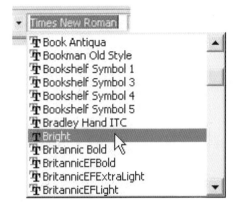

Times New Roman

- 𝐓 Book Antiqua
- 𝐓 Bookman Old Style
- 𝐓 Bookshelf Symbol 1
- 𝐓 Bookshelf Symbol 3
- 𝐓 Bookshelf Symbol 4
- 𝐓 Bookshelf Symbol 5
- 𝐓 Bradley Hand ITC
- 𝐓 Bright
- 𝐓 Britannic Bold
- 𝐓 BritannicEFBold
- 𝐓 BritannicEFExtraLight
- 𝐓 BritannicEFLight

Click and drag the mouse over the text you wish to select.

ABC def 123
ABC def 123
ABC def 123
ABC def 123
ABC def 123
ABC def 123
ABC def 123
ABC def 123
ABC def 123

Not all fonts have a lowercase alphabet.

Choose the font style carefully. Some styles are very ornate and not suitable for the body of your work. The body text (i.e. the main written part) should be a simple, clear font which is easy to read. Times New Roman is ideal. It has serifs (the little bits at the ends of each letter) which is thought to make the font easier to read. If you don't like serifs, use a font like Arial which is a sans serif font (without serifs). The ornate styles like those shown on the left should only be used for the main title and never for body text.

Font size

Fonts are measured in point sizes. You can select the size of the font in the same way as choosing the style by clicking the arrow alongside the current size and choosing the new size from the drop down list. You can also type in a new point size that does not appear on the list if you wish e.g. 9.5.

Sizes vary between fonts. For example, 10pt Arial is about the same size as 11.5pt Times.

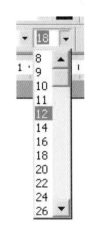

Like font styles, you should use font sizes very carefully. In general, don't use a font size any greater than 14 for the main body of your work. Preferably, smaller sizes should be used. Although 10pt looks small on screen, when printed out it's actually quite readable.

You can use a slightly larger font for any subheadings. There are no firm rules, but a 1 or 2 point increase is usually about right. For the main title, you can use something much larger and/or more ornate if you wish.

Emphasising text

If you want to emphasise a word or phrase you have two options:

Don't use the underline feature which is a throwback to the days of the typewriter. If you want to highlight a phrase there are much better ways of doing it.

1   Select the piece of text you want to emphasise.

2   Click on the B symbol to **embolden** or the I symbol to *italicise.*

Highlighting text

Most word processors allow you to change the text colour, whilst some allow you to highlight sections of text in the same way as you might with a highlighter pen. This effect is a handy way to help with revision but would not normally be used in your homework assignments.

Positioning text

There are four ways in which the text can be aligned on the page.

Left

In most cases it will default to Left-Aligned. That means that lines of text will begin on the left margin, but because each line of text will contain different numbers of characters, the right hand margin will appear rather ragged. This book is Left-Aligned and if you look at the right margin, you'll see what I mean.

Justified

Some people like to fully justify the text. This is where the computer increases the space between each word so that the right margin, like the left, is straight. Most newspapers are fully justified and when you work in columns is does look quite good, although if the text size is large and the columns narrow, you could find that you'll have one word at the left of a line, another word at the right and a huge gap in between. This paragraph is fully justified.

Right

Right-Aligned is the opposite of Left-Aligned and would usually be used only when you want to put a couple of words on the right, and not a full paragraph. For example, you might want to end each piece of work with:

Geoff Preston
15th May 2003

If you want to move text across a line, never use spaces to do it. Use the Tab key instead.

Note how the text lines up on the right. Note also that this is not the same as pressing the space bar dozens of times to move the text to the right. When writing a letter, this is a good way of getting your address at the top right.

Centred

The final alignment is Centre, which as the name implies, centres the text. Although this paragraph is centred, you would normally use it for titles only.

Bullet points

If you have a list of points to make, rather than simply including them in a sentence, you can make them really stand out by bulleting them.

1 Enter the text then select the lines of text you want to bullet.

2 Click on the Bullet icon on the toolbar.

Do not use this feature too many times in the same document.

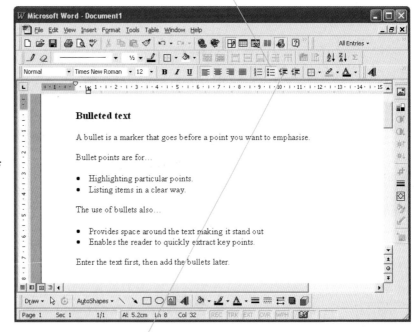

3 If you want to further indent the bullet point, click the Indent icon.

4 If you want to change the bullet style, select Bullets and Numbering from the Format menu.

Inserting objects

If you wish to illustrate a piece of text with a picture or two, you could print out the text, then print out the pictures and, using a pair of scissors and some glue, stick it all together. But there is a much better way. Most word processors will allow you to put a picture into the text before printing.

1 Once you have all of your text entered, click the mouse pointer roughly where you want the picture to go.

2 Open the Insert menu and choose Picture. You will often be asked if you want to choose a picture from the clip-art library or from elsewhere – a picture you have taken yourself, for example.

3 Locate the picture and click the OK or Insert button.

Add a caption to your picture if one is required.

When wrapping text around a picture, do not put the picture in the centre of the page so that it splits a line of text.

You can also insert tables into the text in exactly the same way.

Pictures can generally be inline or floating. Inline means that they will be embedded in the text, whilst floating means that you can move the picture around and have the text flow around it.

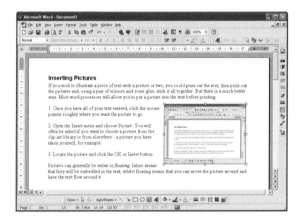

Ensure you have a gap around the picture so that the text does not rest up against the edge.

You can adjust the size of the picture by clicking on it and then dragging one of the handles on the edges and corners of the picture. In most cases, use the corner handles to ensure the picture's aspect ratio (the height to width ratio) remains the same. Further adjustments can be made by right-clicking on the picture and choosing Properties from the menu.

Headers and footers

A header is a piece of text that is automatically printed at the top of each page. Most word processors have this feature and if you're creating a multi-page document it's worth using a header so that your name, the title of the piece, the page number or whatever else you want are automatically placed on each sheet.

A footer is exactly the same except it goes at the foot of the page.

1 In the View menu, choose Header and Footer. A floating toolbar will open. If you use this feature regularly you could include the floating toolbar with the other icons at the top of the window.

To maintain consistency, create a blank document which will form the basis of all your work. When starting a new piece of work, double-click the icon to open it and then choose Save As [new name] so that your template remains blank ready for the next piece of work.

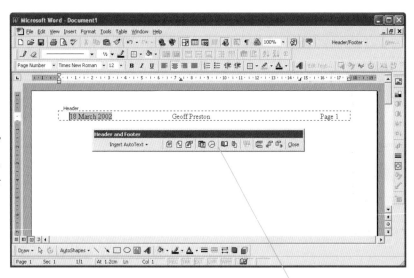

2 Choose from the list of icons on the floating toolbar those features you want to be included in the header.

3 At the bottom of the page you can include the items you want on the footer.

4 When you've finished, click the Close button and carry on typing as before.

Proofing

When you've got your work looking just right, it's time to check and re-check to ensure there are no mistakes.

Before printing your work, carry out a spell check:

If your piece of work contains lots of names, the Spell-Checker may query them. Check the spelling first and then add them to the dictionary.

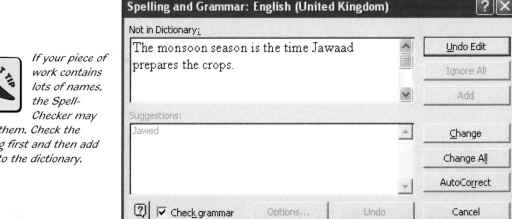

Spelling and Grammar: English (United Kingdom)

Not in Dictionary:

The monsoon season is the time Jawaad prepares the crops.

Suggestions:

Jawed

Undo Edit

Ignore All

Add

Change

Change All

AutoCorrect

☑ Check grammar   Options...   Undo   Cancel

The worst person to read your work is you. If it's really important, get someone else to look through it for you.

This will ensure you have at least spelt every word correctly though that doesn't necessarily mean there are no mistakes. Although you have the correct spellings, there may be incorrect words. You may, for example, have used 'there' for 'their'. Some word processors include a grammar checker but these are not foolproof. You should always read through your work very carefully.

Checklist

- Use correct spellings – especially names and technical terms.

- Check the grammar – words, parts of speech etc.

- Check the factual content – dates etc.

- Ensure you have a title – at the top of the page.

- Ensure you have included your name and date (and possibly your tutor group).

- Ensure you have saved the latest, fully corrected version of your work.

Printing

Although some teachers might be happy to take a pile of floppy disks home for marking, most would prefer paper. That means you'll need to print your work.

Covers

Chapter Six

Hardcopy

The term hardcopy refers to the printed output from a computer program and in most cases this is what you'll need to give to your teacher for marking.

Preview

Check your work thoroughly before printing as mistakes can only be rectified by reprinting.

It was thought that the introduction of computers would reduce the enormous waste of paper produced by businesses and educational establishments. In fact, it's had precisely the opposite effect. Before computers, a typed page would have to be completely retyped to correct anything but the smallest of errors. Now computers have arrived, all you have to do is press a button and another copy can be produced in seconds.

So, before printing you should preview your work.

Most Windows programs provide Preview via an icon at the top of the window or an entry on the File menu

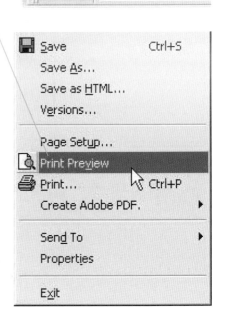

Choosing this option provides you with a view of exactly what the paper will look like when it comes out of the printer.

If you have a choice of printers, make sure you choose the one you want to use before previewing.

Apart from showing any errors in your work, it will also prove that it will fit on the paper you're intending to use without parts missing from the edges.

Setup

Most Windows programs work in the same way and provide three ways of starting a print-run. The first is simply clicking the Printer icon at the top of the application window. In most cases, this will start the printer with the current settings.

Pressing Ctrl+P or choosing Print from the File menu on the menu bar leads to a dialog enabling you to set up your printer.

The exact contents of the Printer dialog will depend on which printer you are currently using.

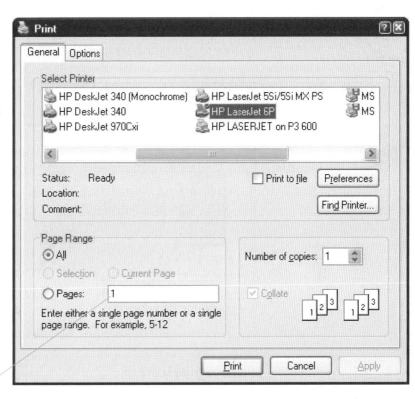

If you find one page contains a mistake, don't reprint the whole document, just print the one page which you can choose from the Print dialog.

Choose the printer you want (if you have more than one) and click on the Preferences button to make changes like setting the paper orientation – whether you want the long edge of the paper horizontal (landscape) or the long edge vertical (portrait).

You can also choose the print quality and the type of paper you intend printing on.

Two-sided printing

To get really professional results for your coursework and major assignments, print your work on both sides of the paper. Some printers have this feature built in and automatically turn the paper for you. If your printer does this, you're lucky. If not, two-sided printing can be achieved with any printer and a little thought.

The basic principle is to print all the odd pages first, then turn the paper over, put it back in the printer and print all the even pages.

Well, that's the theory but depending on which side your printer prints on will determine precisely how the pages are put into the printer to print the backs.

Check which side of the paper your printer prints on.

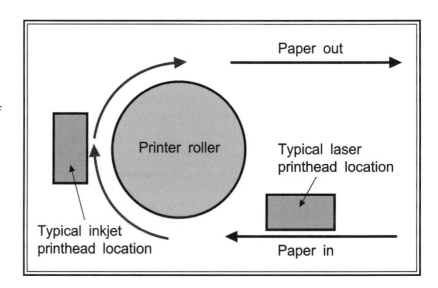

Face up printing

The roller effectively turns the paper over.

Some laser printers print on the side of the paper that is facing up when placed into the paper tray. When the paper comes out of the printer, the printed side will be facing down, unless the printer has a device to send the paper out of the back of the printer.

Face down printing

Most inkjet printers print on the side of the paper that is facing down when placed in the paper tray. When the paper comes out of the printer, the printed side will be facing up.

If the paper comes out of the printer face down

After you've printed the odd pages, take the pile of printed pages out, and turn each individual page over, keeping them in the same order. (Alternatively, turn over each piece of paper as it comes out of the printer.) Put them back into the printer so that page 1 is at the top and facing down, (the side facing up will be blank, ready to have page 2 printed on it) and print the even pages. When the paper has been through the printer twice, you will need to turn over each piece of paper to get them into the correct order.

If your printer has a flap or lever that allows the pages to exit the printer at the rear, activate it before you begin printing. The pages will then exit the printer face up. Take the pile of papers, turn the whole pile over so that page 1 is at the top, but face down. Now print the even pages, again sending them out of the rear of the printer. When you take out the printed pages they will be in the correct order.

If the paper comes out of the printer face up

After you've printed the odd pages, take the pile of printed pages out and turn each individual page over, keeping them in the same order. (Alternatively, turn over each piece of paper as it comes out of the printer.) Put the paper back into the feed tray so that page 1 is at the top, facing up, and print the even pages. After going through the printer twice, the pages will be in the correct order.

If you're working on a multi-page document, put numbers on the pages. It will help you order them correctly.

In all cases, the paper goes into the printer so that the top of each page goes in first.

Some print dialogs provide options for printing odd and even pages. Others force you to resort to entering each of the page numbers you want printed.

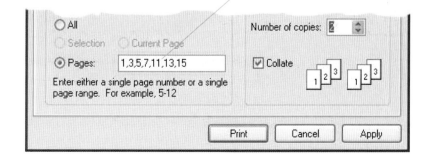

Paper

Choosing the correct paper for your printer and for the way you intend printing is very important.

If you use a laser printer, you can put almost any printer paper in it, although 80gms/m^2 laser paper or photocopy paper will give adequate results.

For inkjet printers, things are not quite so simple. Although you can print onto cheap photocopy paper, the results are not usually very good, especially if you select the Best Quality option from the Print Properties dialog. Certainly don't attempt to print on both sides of photocopy-type paper with an inkjet printer as the paper won't be able to absorb all the ink and will get waterlogged.

The thickness of paper is described by how much a square metre would weigh. The more a piece of paper weighs, the thicker it is.

In almost all cases, you're better off using paper which has been manufactured especially for inkjet printers. If you want to print on both sides, buy paper which is designated as suitable for double-sided printing. For printing mainly text with a few pictures, use 'Normal' quality printing on inkjet paper of 80–100 gms/m^2 thickness.

If you're printing out work which is mainly pictures (i.e. for Art), use thicker 140gms/m^2 or even 190gms/m^2 which is thicker still.

Submitting homework

Once you've completed your work, you need to prepare it for submission to your teacher.

Covers

Chapter Seven

Multiple pages

The first thing to check is the date you are to hand in your homework or your coursework assignment. Before you hand it in, make sure your name and tutor group are clearly printed on each page.

If you have lots of pages to give in you should fix them together in some way. Possible methods of fixing are:

You may only have one piece of work to complete but your teacher will have 20–30 pieces to mark.

Staples. Using a staple at the top left corner makes the document look like a throwaway pamphlet. Two or three staples down the left side is preferable but take care that the staples are in line and that the pages are flat.

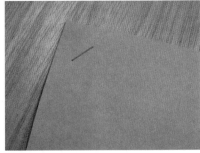

Comb binder. The cost of these machines has fallen rapidly and is now a realistic option for home use. There are two processes involved in fitting the comb. The first uses a machine that punches rectangular holes in the pages. The second is the machine that opens the comb so you can lay the pages onto the prongs before closing the comb to bind the pages. This makes a very professional-looking document.

Ensure your work is handed in on time if you do not want to risk losing marks.

Folding and tearing the top corner. This is the worst way of attempting to fix pages together. Don't ever use this method. It's horrible.

Plastic spine. You can buy a box of plastic spines from most good stationers. You simply slide the spine over the pages and it's done. Some kits include transparent leaves which provide protection for the front and back pages. You can buy spines in different widths to cope with different numbers of pages, but in all cases this method seems to work best when you're binding more than ten pages.

Transparent envelope. This is quite a good method for a small number of sheets – say up to five.

If you intend punching holes into pages, set your word processor with a wider margin so you don't punch holes through the text.

Cathedral tags. These are short pieces of string with a plastic toggle on each end. Using a hole punch, punch holes in the margin and thread the tags through the holes in the pages. This method is more suited to a few pages.

Bulldog clip. This is another device to avoid. A bulldog clip is great for use with a clip board when you're out collecting survey data, but that's where it should stay.

Paper clips. These are good in theory, but in practice they are too temporary. If you have half-a-dozen projects held together with a paper clip, pages from one project soon begin migrating to another.

Moving softcopy

Softcopy is the name given to data (your work) which is saved on the computer's hard disk or on other computer-readable medium like a floppy disk. (The hardcopy is the printout.) If you use the same computers and the same software at home and school, you could transport your work to school and print it there. Additionally, a piece of work that was started at school could be taken home so that it may be continued there and returned back to school when complete.

There are three ways of moving softcopy:

Email. If you have an email account at home and another at school, you can email the file(s) to yourself.

If you do not have two email accounts, send the message to yourself but click the Send button and NOT the Send and Receive option.

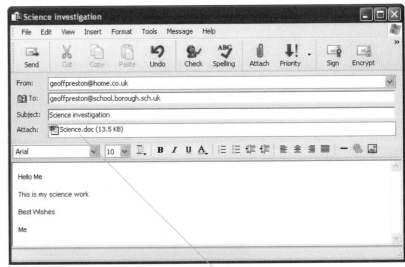

Although any email address can be accessed from anywhere in the world, in practice many users access their 'school' email address from school and their 'home' email address from home.

You can attach a file to an email by simply dragging it into the Create New Message window. It will appear in a new pane labelled Attach.

When you get to school, download your email and you will have your attachment which you should save before attempting to work on it.

Sending the file back home is a reversal of the process: send an email to your home account and collect it when you get home.

Floppy disks are quite delicate. Carry them in a protective case and take care not to let them come into contact with dust or dirt.

If you intend using the CD method, use re-writeable disks which can be used over and over again.

Don't leave the disk in the school computer.

Disk. You could copy your work onto a floppy disk and take the disk from school to home (or vice versa). If you're going to try this, take care of the disk and ensure you have copies of your work on the computer on which you began your work *and* the computer on which you added to your work.

Floppy disk drives are not very reliable after they have been used by hundreds of people and it's quite common to find a drive that reports a disk fault when another drive happily reads the disk.

A variation on this theme is to use a CD writer – a device that will write onto a special CD. After you've copied your work onto it, the disk can be formatted so that it can be read by a standard CD ROM drive. If you want to take your work into school to print, you'll need a CD writer at home. If you want to take work back and forth between home and school, you'll need one in both places.

Whichever method of transferring files you choose, you must have a current virus protection application at home. Your school should be running virus protection software but it is your responsibility to ensure you don't get a computer virus. Remember it is likely you won't be the only student moving work in and out of school.

Computer. If you're lucky to own a laptop computer you could use that to take the work into school. The best laptop computers have an infrared eye to enable you to connect to other infrared devices. This method avoids the need for cumbersome cables. If your school has a computer with an infrared eye you should be able to copy the files onto the school computer. If you just want to print your work, see if the school has got a printer with an infrared eye.

Converting files

Life would be so much easier if there was just one of everything. If there was just one type of computer, running one operating system and there was only one word processor, one spreadsheet and one painting program, we could all be assured that a file created on one computer could be read by every other computer in the world.

Unfortunately, having two computer standards (Mac and PC), several operating systems (Windows, Linux, Mac OS) and a huge range of software by countless software companies, the chances of compatibility between computers is remote. Unless, that is, you have specifically bought the same computer and the same software that your school uses.

In reality, that is not always going to be the case, although many of your files will have a degree of compatibility if you have the same computer and the same operating system.

For those who do not, you're going to have to work a little harder if you want to move your work to and from school.

These instructions assume you have started a piece of work at home and want to take it to school. It works in exactly the same way if you're trying to move work created at school to home – just do at home what the instructions say to do at school, and do at school what the instructions say to do at home.

General guidance

If you create a piece of work in one word processor, there's no guarantee that it will be able to be read by a different word processor. If this is the case and you need to move work between home and school, you will need to save your work in a different file format so that it can be read by another, similar program.

1 On the school computer, open the program that you want to be able to read your work.

2 Create a new document.

3 Choose Save As... from the File menu.

Always save your work in the usual way to begin with. This will ensure you have a version you can read on the computer on which it was created.

Some programs use a separate entry on the File menu called Export. Check for supported file types there.

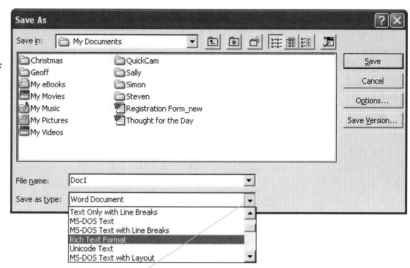

4 Click on the arrow alongside Save as type and make a note of all the file types that that program will read. Many will probably be older versions of the same program (e.g. Microsoft Word 97 supports Word 6, Word 95, Works 2 and Works 3).

5 Check with your program at home to see which file formats supported by the school program are also supported at home.

If there is a common format, save your work in that format at home and transport it to school. If the software uses Import, then choose that option to load the work done at home into the school computer.

If you have difficulty getting the file created at home into the program you want to use at school, try opening a new document and 'dragging' the file into it.

Word processed files

Files created in a word processor are usually the easiest to move, providing they don't contain pictures.

File formats that virtually all word processors can read are Text and RTF. Text, or ASCII Text as it is sometimes called, is just the words and nothing else. No page sizes, no font information, no paragraph styles. This is really a last resort. It is almost guaranteed to work, but because you're only moving the text and nothing else, you could still be left with a great deal of work to do.

Rich Text Format (RTF) is better because it does attempt to include information about the document including font styles. The problem here may be that if you've typed your work in a font style the school computer doesn't have, you'll still need to do some layout work.

Test this out as soon as possible – but certainly before you actually **need** *to do it.*

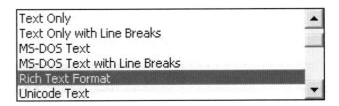

Word processed files that contain pictures are more uncertain. Some word processors will save files in a format specifically used by another word processor e.g. StarOffice will output Microsoft Word files. If you can find that sort of match, you should be home and dry. But before throwing a party, check that it actually works. Some files created in Microsoft Word can sometimes lose some of their formatting when opened in StarOffice.

If you've got pictures in your work it's often best to save them separately and re-insert them when you've got the word processed file opened at school.

Spreadsheet files

Like word processors, you can usually move data between different spreadsheets, but it's not always an exact science.

There are two file formats that will work to a limited extent: CSV and TSV. Comma Separated Values and Tab Separated Values (sometimes called Comma or Tab delimited files) will save the numbers from your spreadsheet, with either a comma or a tab separating each cell on a line, and a return character at the end of each line:

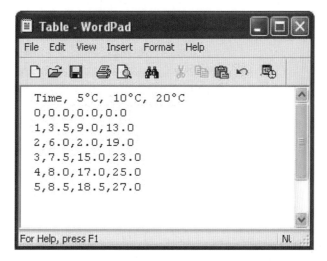

Using methods like CSV and Plain text almost always means you'll lose some of the layout.

Most spreadsheets can read CSV or TSV or both, but all you will save are the basic numbers. Any cell formatting, including colours and grid lines, will be lost. More importantly, it won't save the formulae, only the results of the calculations – the numbers.

Database

It can be quite tricky to move database files from one database program to another. CSV or TSV are often the only options available to you and once again, any clever layout work will be lost.

Vector graphics

There are several vector graphics file formats and rarely do programs seem to include many (if any) of the variations. But don't despair. Most applications of this type can export and import WMF – Windows Metafiles. WMF is the 'standard' file format used in Windows and files usually transport between different vector applications quite well. The problem comes when a particular graphics application can do something that Windows Metafiles cannot. When this happens it usually means that that particular part of the drawing is simply omitted.

Bitmapped graphics

I've lost count of the number of file formats there are for bitmapped graphics, but the main ones are BMP (Windows Bitmap), JPG (or JPEG – Joint Photographic Experts Group), TIFF (Tagged Image File Format), GIF (Graphics Interchange Format) and PNG (Portable Network Graphics).

It seems that most bitmap applications will read most, if not all, of these. The problem comes when the bitmapped picture contains more information than the picture itself. Many digital photograph editing applications feature a range of effects like splitting parts of the picture onto different layers, or saving the history of the alterations. These files are unique to each individual application and therefore cannot usually be read by an application other than the one used to create it. The only possibility you have is to export the picture as a JPG, BMP etc. in which case layers and history will be lost.

Different computers

Moving files between different computers shouldn't present much more of a problem, as many programs running on a PC have Mac equivalents with identical file formats. The main problem is likely to be getting one computer to read a disk that was created on a different computer. Apart from the very earliest examples, Apple Macintosh computers can read PC-formatted floppy disks, though not necessarily the other way round.

In short, if you save your work on a PC-formatted disk you should be able to move files from PC to Mac and back again. Whether you can read the files once you've got them there is another question altogether.

Subjects

This chapter explains how computer skills learned whilst working for one subject can often be of use in another.

Covers

Chapter Eight

When to use a computer

During your first years at secondary school you will study about a dozen subjects and most of them will involve some homework. Some pieces of work will be short and will therefore be completed quite quickly. Other types of homework may be longer pieces which could take many hours spread over, possibly, several weeks.

Of all the homework assignments you will receive, many could be completed with the help of your computer although you will not necessarily be able to complete all of your homework in this way. In fact, it is not a good idea to try to complete all of your homework using your computer as there are other skills that should not be ignored but developed, like handwriting and drawing.

The trick is to learn when a computer is appropriate and when other, more traditional methods should be used.

No ICT?

One of the subjects you'll be studying at school is ICT – Information and Communications Technology – yet this book lacks a separate section on it. That's because ICT is part of most other subjects. You will improve your ICT skills by completing a range of homework assignments for your other subjects using your computer. It's rather like getting something for nothing – by completing your homework using your computer you learn more about the subject you are studying *and* improve your ICT skills.

Check first

Your teacher may not want a particular piece of homework completed using a computer. Check first.

Worksheets which require you to work on the actual sheet are generally not suitable for computerised homework. It is usually the longer pieces of work, sometimes referred to as project work, which require you to present several pages of writing illustrated with drawings, diagrams and photographs that are best suited to computer output. If in doubt, ask your teacher if you can complete a particular piece of work using your computer.

What's in this book?

The remainder of this book is divided into subjects, with each subject being divided into (usually) three sections:

Example homework

It is impossible to predict every piece of work you are likely to be given and provide an explanation of how it could be completed using ICT. What this book tries to do is give some ideas of how the computer could help with pieces of homework that you could be given for each subject. At the very least, each piece shown will be of use for your subjects and should also improve your ICT skills.

Websites

Research will often be part of your homework. A few really good sites have been included in each chapter to help with researching. These sites are in addition to those listed on pages 47–48 which offer general help for most or all subjects.

No homework?

There will be times when you have no homework or complete it quickly. These sections provide some ideas for you to extend your homework in a way that should also extend your ICT skills.

Transferrable skills

Information & Communications Technology is not just a subject on its own, but a tool to enable you to work more effectively and more efficiently. That means in *all* subjects.

What is interesting about ICT, and is probably unique, is that the ICT skills you learn and develop whilst working for one subject are almost always transferrable to another subject. Sometimes the skills are transferrable to several subjects and occasionally all subjects.

Always look for ways to improve your ICT skills as well as your subject skills.

When you've completed a piece of work with your computer, ask yourself what ICT skills you have used that can be transferred to other subjects. For example, if you've learnt how to create a graph for, say, Mathematics, ask yourself where else those skills might be used. If you haven't discovered it yet, you'll soon find that graphs also appear in Geography and Science. The purpose for each graph may be slightly different for each subject, and the type of graph may be different, but the basic skills needed to create the graph are exactly the same.

The nature of ICT is that skills are transferrable between subjects, although in some cases you may need to adapt the skills slightly when applying them to a different piece of work.

The diagram on the inside front cover shows some examples of where different types of applications can be used to produce work for different subjects. This should help you apply ICT to the different subjects you study and help you recognise how the skills from one subject might be transferred to another.

So when you're next creating a piece of work for technology using a drawing program, it might be worth looking at how that program could be used in, say, Science.

Art and Design

You can use a computer to imitate traditional painting techniques, but it's often better to use techniques which cannot easily be reproduced without a computer.

Covers

Chapter Nine

Wallpaper

Copying parts of a picture is something a computer does really well which means if you are asked to produce a repeating pattern, like wallpaper or a fabric design, a computer is likely to be better than traditional art tools.

If you're using a vector program, select all of the elements of the pattern and group them using (usually) Ctrl+G.

You can use either a vector or a bitmapped graphics program to do this. Create the repeating pattern (in this case, a flower), then:

1 Mark the pattern by choosing the Selection tool and dragging a rectangle over the shape. Copy it with Ctrl+C.

2 Use Ctrl+V to paste a copy back into the picture.

3 Drag the copy into position.

Use Copy whenever possible. The flower was created from just 1 ellipse, 1 circle and 1 line. Having drawn 1 yellow petal, copy it, rotate it 45°, copy, rotate etc. The leaves were copied from a petal, and changed to green.

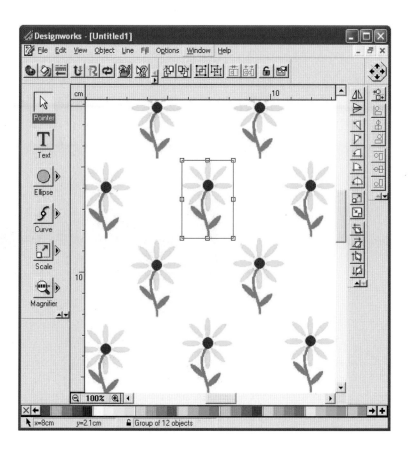

Digital pictures

The increased affordability of scanners and digital cameras means that many more people can produce digital images and consequently use them as the basis of some interesting artwork.

Many cameras and scanners are supplied with software which will enable you to make major alterations to the pictures beyond simply cropping to size. If you haven't got one of these there are free or trial versions available on the Internet (try www.ulead.com/) and very cheap examples on sale in many high street stores.

The most advanced digital imaging techniques involve 'cutting' a piece from one picture and either pasting it into another or creating an effect with the remaining part before replacing the part you first removed.

There are many simpler effects that can be applied to an image which will produce effects that could not easily be achieved any other way.

I've selected just four imaging techniques which are found in most programs.

Although these pictures were taken with a digital camera, you can get similar results by scanning photographs or downloading pictures from the Web.

A pixel is the shortened form of 'picture element' and is the smallest dot on the screen that can be individually coloured.

This technique was used to great effect by Andy Warhol, although he didn't have the benefit of having a computer to do the hard work. I've applied a threshold to the picture on the left so that every pixel above a certain brightness in converted to white, and all pixels below that level become black. It's interesting to note how much detail you can remove, before the picture becomes unrecognisable.

Having got your picture in just two colours you can then experiment with colours. This is easily done in Paint.

Open the picture in Paint and enlarge the work area by dragging the small marker at the bottom right of the picture.

If you are using Paint, when positioning your copy, select the Transparent button at the bottom of the toolbox.

1   Make a copy of the image by choosing the Select tool on the top right of the toolbox and pressing Ctrl+C.

2   Use Ctrl+V to paste a copy back into the picture.

3   Drag the copy into an empty part of the work area.

You may need to magnify the picture to fill any small areas that are not connected to the main area being filled.

4   Select a colour from the palette and use the Paintpot tool to fill one area. Change colour and fill the other area.

5   Select the recoloured copy and drag it into position alongside the first picture. Continue steps 1–4 until you have built up a collection.

The Andy Warhol Museum website is at www.warhol.org/.

There are many distortion methods featured in digital imaging software, including some that attempt to correct exaggerated perspective.

There are many digital imaging applications and they all feature a different armoury of tools, but most feature those shown here.

This effect is reminiscent of a piece of work by the Dutch artist Maurits Escher who drew a self-portrait whilst looking into a crystal globe. The Official M C Escher website shows all of the artist's work, including the self-portrait in a globe, and may be found by visiting www.mcescher.com/.

The finished picture (above right) has actually had two processes applied to it. First, the centre of the image was distorted into a sphere. When this option is chosen, the software will create the largest possible ellipse within the rectangular perimeter of the picture. The only way you'll get a perfect circle is to begin with a picture that is perfectly square. If one side of the picture is slightly longer, then the result will be an ellipse unless the software offers you a degree of control over the precise shape.

Always keep the original picture in case you need it again.

Once the distortion effect has been applied the picture can remain in that form, but I think it's better to remove the part of the picture which is outside the distorted area. For this you should select an elliptical cropping tool and draw the largest ellipse you can within the picture. If the ellipse is drawn carefully so that it just touches each of the four sides, then you should have exactly the same size ellipse as the distorted area.

You should note that you will have selected the part of the picture you wish to keep and so you will have to Invert the selection before deleting.

Changing the colours can produce interesting effects. This particular process is referred to in the software as 'Electrify' although similar effects appear in different programs with other names. Not only can you control the colours that are used to replace existing colours, but you can determine how many similar shades are replaced by a single colour, thus reducing the number of colours that appear in the final picture. This effect is regularly seen in advertising.

Pixelating is an effect that one usually tries to avoid. It happens when a bitmapped picture is enlarged and the result is a very chunky appearance. Some software actually features an effect based on pixelating:

This effect is seen a great deal on television to hide someone's identity.

 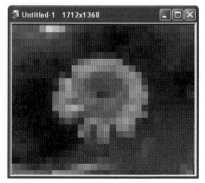

You can apply the effect to one part of the picture (in which case you select the area you want to pixelate in the same way as the 'car in the sphere') or you can apply it to the whole picture. Either way, you can usually control both the size and shape of the pixels.

Websites

Most art galleries have their own website featuring pictures of the exhibits and some information about both the exhibit and the artist. The Louvre (www.louvre.fr/) is typical: there are regular features and a guide to special exhibitions and guided tours.

 If you enter the name of an art gallery into the address bar or search panel you'll almost always find a website.

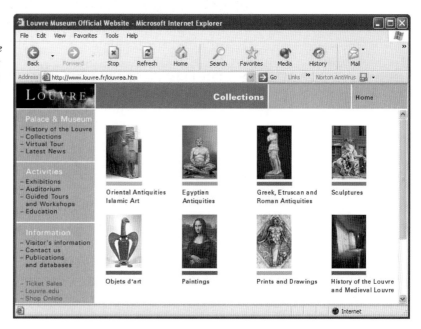

Artcyclopedia

This huge resource at www.artcyclopedia.com/ will be invaluable to History of Art students. You can search for an artist, work of art or art movement. So, if you're trying to find out about Northern Renaissance, this should be your starting point.

Art Art Art

This online art gallery of sculpture, ceramics, film, glass, and jewellery is at www.artartart.net/. The site showcases the work of several modern artists. Most pieces on display contain a link so you can email the artist.

ArtZone – BBC Online

This popular BBC site at www.bbc.co.uk/art/ features news and articles from the world of art. There are also links to lots of resources.

No homework?

Apart from easily copying elements of a picture, you can also rotate elements. Draw a square to represent a ceramic tile and on it draw a simple pattern. By copying and rotating tiles you can create different patterns.

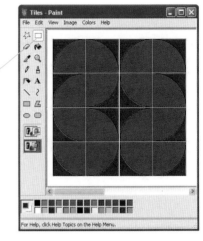

1 Begin by drawing a single tile and add a pattern to it.

2 Select the tile by choosing the Select tool from the top right of the toolbox and drag a rectangle over the tile starting at the top left towards the bottom right.

This exercise was done using Paint, but a similar result could be obtained from a vector drawing program .

3 Make a copy (Ctrl+C).

4 Paste it (Ctrl+V) – the copy will appear in the top left of the window.

5 Open the Image menu, choose Flip/Rotate and turn the selection by either 90°, 180° or 270°.

6 Drag the rotated tile into position, alongside one of the other tiles.

Repeat steps 4–6 until you have 16 tiles arranged in a 4x4 square.

Now rearrange some of the tiles to change the pattern made by the block of 16 tiles. Simply select a tile, rotate it and carefully reposition it.

See how many patterns you can get from 16 tiles. There's a lot more than you think.

Design and Technology

You can use your computer to do a great deal of design work for this subject.

Covers

Chapter Ten

Drawing

There's not much prospect of any computerised manufacturing at home unless you have some expensive control hardware, but if you have some Lego, you could save up for a control interface which would enable you to control your Lego model (which could be a buggy, robot arm or almost anything that has a motor or two). For more information about Lego control visit the Lego website at www.lego.com/.

It is more likely that elements of the design process, particularly final working drawings, will be completed by computer.

Working drawing

Most vector drawing applications allow you to construct both 2D and 3D drawings by changing the type of grid.

The 2D grid is the usual format and provides dots in rows and columns enabling you to draw vertical and horizontal lines. It is in this mode that you would draw your final working scale drawing as an orthogonal projection.

When setting the grid, ensure you lock or snap to the grid. Grid settings are usually in the View menu.

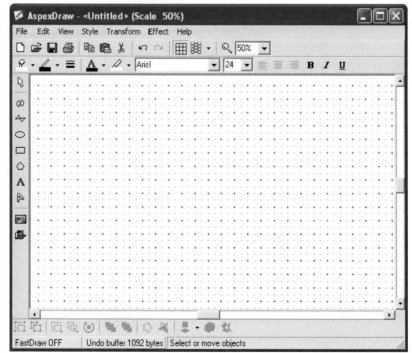

There are two methods of Orthogonal drawing known as 1st Angle and 3rd Angle. 1st Angle, once the favoured method in Europe, is now used less frequently as the American 3rd Angle Projection is gradually taking over.

If you want to construct an orthogonal drawing using your computer you should really use a vector drawing application and not a bitmap drawing program. The advantage of using a vector drawing program is that the final drawing can be output using a plotter. A plotter is a tool you're unlikely to have a home but your school will probably have one. It uses a pen on a movable arm which will draw your work almost exactly the same way as you would construct a line drawing although it does it much faster.

If you only need to show one side elevation, divide the page into 4.

Having established the size of paper, divide the page into six (three across and two down). If drawing in 1st Angle, put the Plan (top view) in the bottom centre area, the Front Elevation above the Plan, the Right Elevation to the left of the Front Elevation and the Left Elevation to the right of the Front Elevation.

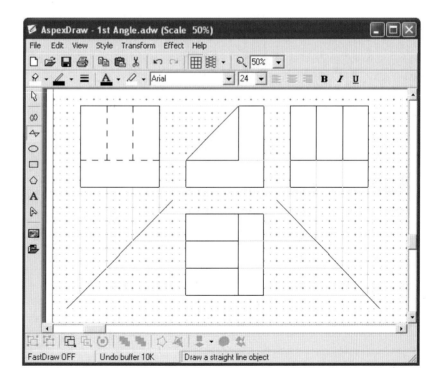

For 3rd Angle Projection, the Plan goes at the top, the Front Elevation underneath the Plan, the Left Elevation to the left of the Front Elevation and the Right Elevation to the right of the Front Elevation.

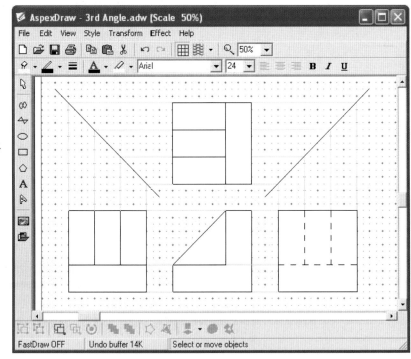

If the drawing contains circles, draw in the centre lines first using a chain line (long and short dashes).

Hidden detail, if required, should be drawn as a dotted line.

If this is to be a true projection, lines are projected to transfer dimensions from one elevation to another. It is common draughting practice to put 45° lines alongside the Plan and pointing to the Front Elevation so that dimensions from the Plan can be projected around a 90° corner to the Side Elevation(s) and vice versa. The projection lines should be drawn either thinner (0.3 mm) or in a lighter colour such as grey. The outline of the elevations should be black and 0.7 mm thick.

Most drawing programs also let you draw lines with arrow heads at one end or both ends which are used for dimensioning a drawing.

To add dimensions:

1 Use thin (0.3 mm) black lines, in line with, but not touching, the outline. These are called datum lines.

2 Add a continuous 0.7 mm dimension line with a long slender arrow head at each end.

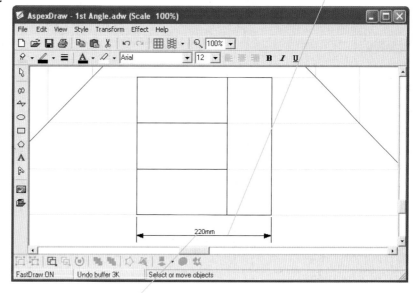

3 Apply the dimension in a sans serif font (Arial is a good choice) above and in the centre of the dimension line.

Isometric drawing

Drawing in 3 dimensions is also possible by changing to an Isometric grid. This grid has points arranged in vertical lines, but instead of horizontal rows of points, is has rows at 30°.

Unless the object you are drawing has angled surfaces, all lines in an Isometric drawing will be either vertical or at 30°.

Begin by drawing as much of one face as you can before moving on to other faces.

Isometric drawing is a method of drawing to scale whilst also showing all sides, but you will never see an object in Isometric.

Try to draw closed shapes rather than single lines so that you can easily add colour to the faces later.

There is another 3D drawing method known as Oblique which basically involves drawing a 2D elevation and then converting it to 3D by adding lines at 45°.

1 Having set the isometric grid, draw one face as a continuous shape so that, if required, it can be coloured.

2 Add further faces to build up the shape.

3 The last parts to be drawn are any sloping faces such as the web shown in this casting.

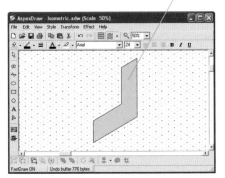

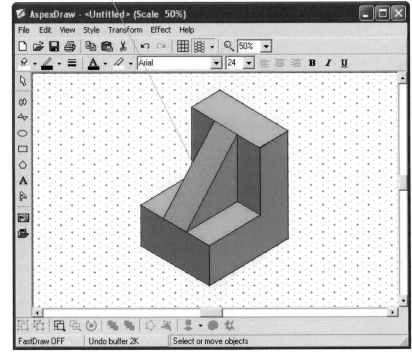

You can, if you wish, add dimensions to an isometric drawing. Follow the steps on the previous page, but remember to keep datum lines and dimension lines at 30°.

Websites

One of the best websites for Design & Technology is Technology Links at www.technologylinks.org/.

This site has links to many other websites covering virtually every aspect of Design & Technology. There is a comprehensive index so you should be able to find exactly what you're looking for very quickly.

The topics include Museums of Technology, Electronics, Mechanisms, Energy, Computer Control, Structures, Ergonomics, Materials & Processes, Tools & Machines, Robotics, Inventors, Graphics, CAD and many, many more.

There are lots of quizzes and worksheets for almost every aspect of this subject.

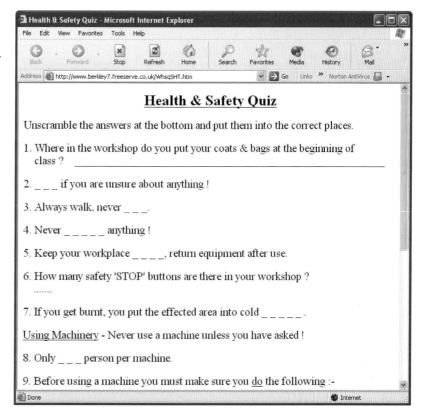

If you prefer you can type the topic you're studying into the Search panel at the top of the screen. You can choose to either search the site, or the whole Web.

No homework?

Perspective

This is the hardest construction to do, but if you get it right you'll end up with a very realistic drawing. Basically there are two methods – single point and two point perspective. (There are a few occasions when more points can be used, but these are quite rare.) Both methods require that you lay down a few guidelines which you can either leave as feint lines (e.g. light grey) or remove altogether.

Single point perspective

This is suitable for drawing either the inside of a room or a scene looking down a street or railway line. As it is unlikely you'll want to output this picture to a plotter, a bitmap drawing program like Paint will be quite suitable. For a room:

Vertical lines will remain vertical regardless of which wall you draw on.

1 Draw a rectangle representing the internal wall viewed as if you were facing it square on.

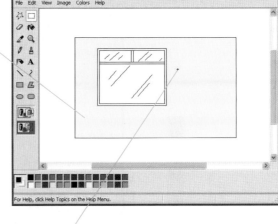

2 Draw any objects on the wall such as a door or windows.

This exercise has been completed using Paint, but it also works with a vector drawing program.

3 On the wall, plot a vanishing point which would be at approximately your eye level.

At this stage, change line width to the thinnest available and/or choose a light grey line colour.

4 Draw lines from the vanishing point through the corners of the wall.

Perspective drawing attempts to take into account the fact that the further objects are away from you, the smaller they appear to be.

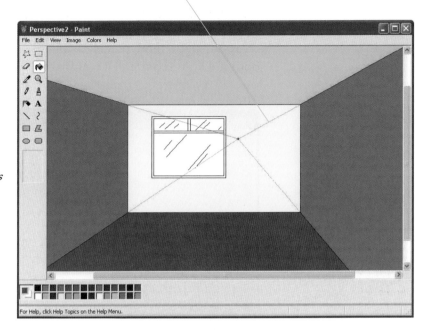

5 When you've drawn the construction lines, draw over them in a darker colour and/or remove the construction lines.

In addition to the wall previously drawn, you will now have a floor, two walls and a ceiling.

Once you've applied some colour to the surfaces using a flood fill, (the Paintpot icon in Paint) the room immediately looks like a room in which you are standing.

Before filling the areas with colour, ensure the lines which outline the walls, ceiling and floor actually meet the edges of the drawing area. If they don't the fill colour will 'spill' into the other areas.

All objects drawn on the side walls, ceiling and floor will be drawn towards the vanishing point.

1 Draw the nearest face first using horizontal and vertical lines.

2 Draw lines from the corners to the vanishing point.

This type of drawing is sometimes called estimated perspective because you have to guess the distances which are going away from you towards the vanishing point.

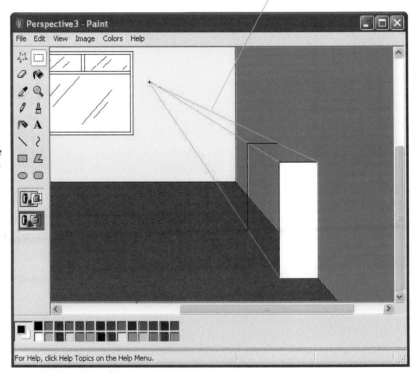

3 You must now estimate how far the furthest face is away from the nearest face and draw the corner with a vertical line.

4 Remove any construction lines and add colour to your object as required.

As you can see from the finished drawing opposite, once you've added detail you can end up with a very realistic drawing.

The more detail you add, the more realistic it will look.

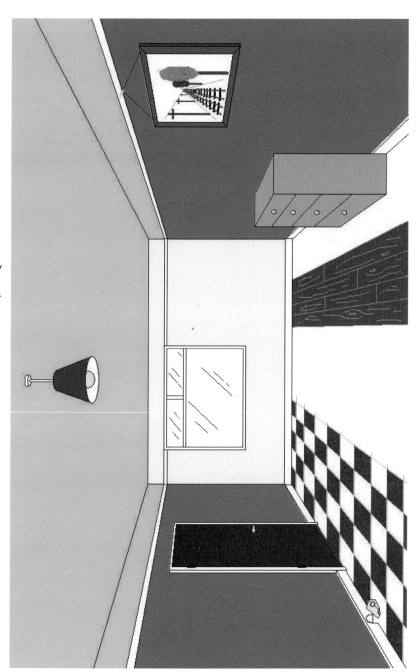

Two point perspective

This is more suitable for small objects that you might make and it's an ideal method of drawing to create a realistic impression of what the object might look like.

2 Draw the nearest vertical corner of your object.

1 Draw a horizon and place a vanishing point at each end.

Always draw the vanishing points as far apart as possible.

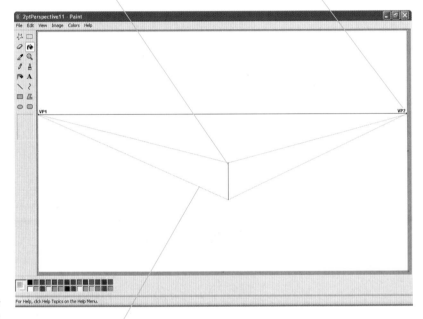

Do not draw the nearest corner too far below the horizon as this will make the object look as large as a building.

3 From the bottom of the nearest edge, draw lines to each vanishing point.

4 Do the same from the top of the nearest edge.

5 Now estimate the length and width of the object and place a vertical line at each point.

6 Where the left vertical line cuts the line to the left vanishing point, draw a line to the right vanishing point.

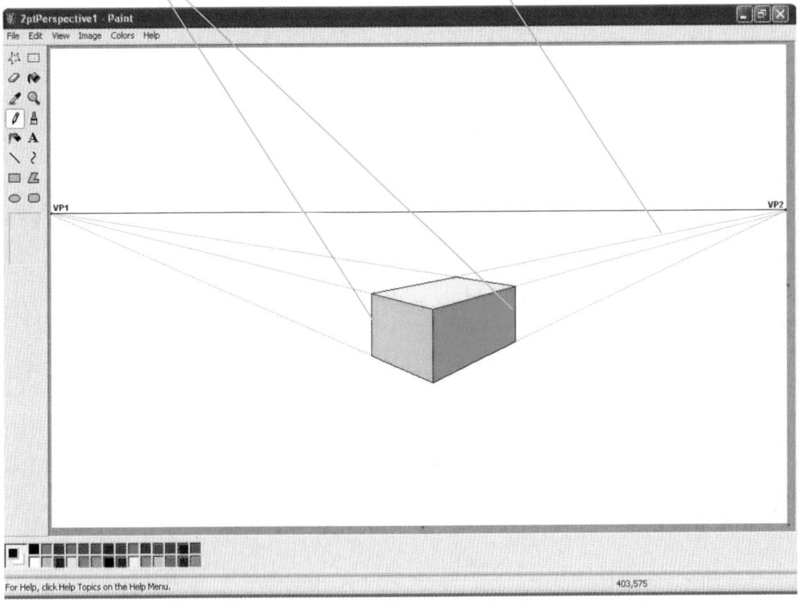

When adding details, apart from sloping surfaces, all lines will be either vertical or will go to one or other vanishing point.

The same construction would apply if you were drawing an object above the horizon (i.e. looking up at the object so that you would see the underside). An object drawn on the horizon would show only two sides because you would be unable to see either the top or the underside.

7 Where the right vertical line cuts the line to the right vanishing point, draw a line to the left vanishing point.

8 Remove all construction lines and/or darken the outlines.

You will now have a box drawn in perspective to which you can add details. For example, this simple box could be detailed to look like a kitchen sink unit or a box to hold chocolates.

Cutting List

Before beginning construction of your piece of design work, (the stage usually referred to as realisation) you should create a list of parts. One of the best ways to create a Cutting List is using a spreadsheet.

Typical headings for the Cutting List are:

Part No. This might relate to the working drawing.

Name of part. It's sometimes convenient to give a name or description to each part.

You could also use a spreadsheet in this way to create a list of ingredients for a recipe.

NoOff. You may require more than 1 identical component to make your product.

Material. This might be a simple generic term like 'wood' although frequently it is better to be more specific e.g. mahogany.

Dimensions. You should enter the finished dimensions of each component. Sometimes it looks neater if you put each dimension in its own column headed Length, Width and Height.

Notes. Here you might want to add any notes that are not apparent from the drawing like '3mm groove along inside face'.

You don't have to use all of these headings.

Cost. You may be required to provide an estimate of the total cost of materials. In the cell at the bottom of the column insert a formula along the lines of =SUM(G2:G10) to add the cost of each item together. (The example formula assumes the Cost column is G, the first row is 2 and there are 9 items in the Cutting List.)

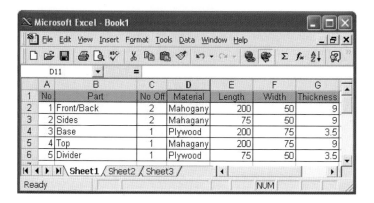

No	Part	No Off	Material	Length	Width	Thickness
1	Front/Back	2	Mahogany	200	50	9
2	Sides	2	Mahagany	75	50	9
3	Base	1	Plywood	200	75	3.5
4	Top	1	Mahagany	200	75	9
5	Divider	1	Plywood	75	50	3.5

Drama

You can do some valuable work for your Drama studies with just a word processor.

Covers

Chapter Eleven

Writing a play

When writing a play, it's common practice to put the names of the characters in the margin and the spoken lines alongside.

To setup your word processor to do this:

| Display the top ruler.

2 Move the Left Margin marker towards the right so that you have enough space to write the longest character name between the left of the page and the Left Margin marker.

3 Drag a Tab marker to the same position as the Left Margin marker.

To write your play...

| Type the name of the character.

2 Press the Tab key and the cursor will jump to the right.

You could put each character's name in bold or in a different font to make it stand out.

3 Type the words that are spoken by that character.

4 At the end of that character's speaking part, press Return. The cursor will return to the extreme left of the page ready for you to enter another character's name.

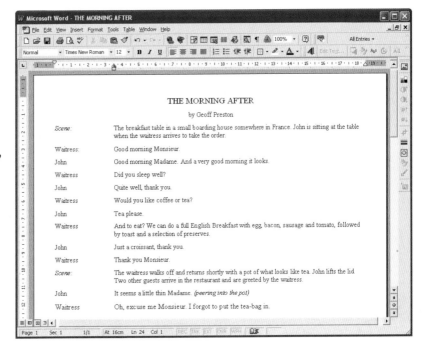

 Don't work too close to the edge of the page. Leave at least a 15mm margin.

An improvement on this is to increase the space after pressing Return so that there is a slight but noticeable gap between each of the character's lines (as shown above). This adjustment is usually found in the Format menu under the heading paragraph. Increasing the size to 1.5 lines is usually about right.

Scene information can be entered in the same way, although you could create a text box which spans across the page and which can be tinted to highlight it.

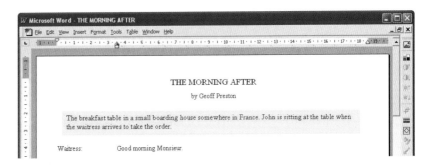

Websites

There's no shortage of websites to help you with all aspects of your dramatic studies.

Typing the name of almost any playwright into the address bar of your browser or search engine will generate a list of websites. Some sites are dedicated to a particular author and contain all the words from the plays with a search facility to help you find the exact quote you are looking for.

If you're not sure where to begin, try visiting Playwrites at `www.playwrites.net/` which features articles and interviews with famous playwrights.

Use the links on Web pages to navigate to other sites.

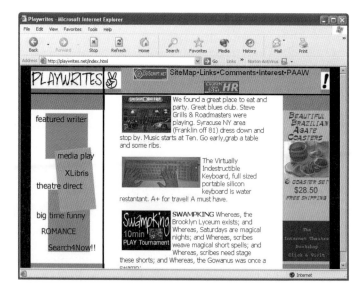

If you need some assistance with writing a play or a scene from a play, visit Learn to Write Drama at `www.successwrite.com/`. You'll find tutorials which will teach you how to write.

The Actors Store at `www.theactorsstore.com/` is an online shop and resource for actors and dancers. You can buy books, dancewear and stage make-up.

Stageplays at `www.stageplays.com/` is an online shop that offers just about every play published in the English language. Use the index to find plays listed by genre, biographies of authors and actors from stage, screen and TV and ticket information.

English

When writing an essay using a word processor, you don't have to start at the beginning and finish at the end.

Covers

Chapter Twelve

Writing an essay

One of the tasks you will be set in English is composition. This may take the form of writing a piece of fiction, a piece of writing to provide information or to convey an argument. It might be to review a book, a play or even a computer game. Whatever the subject matter, a word processor can substantially reduce the workload.

Planning

Your teacher may have a preferred set of headings for planning a composition. If so, use those.

It's not a good idea to just sit down and write (or type). You need to plan your composition very carefully. Begin by starting your word processor and opening a new document. Then use the most appropriate of these headings to help you plan your composition:

Genre. What type of composition is it? A play? A news report? A review?

Setting or Location. Not all compositions will require this, but if you're writing a story it's a good place to begin. You probably won't go into too much detail but ensure you are clear about where the events happen, especially if the location has a bearing on the story.

Incident. If you're writing a story in the form of a news report, you will need to decide what incident is being reported. What actually happened?

Characters. These could be the characters in your story or the main people in your news report. Sometimes it's important to decide certain details about them, like their age and ethnicity. If you are writing a review about a product, find out as much about it as possible.

A word processor is the ideal tool for compositions as you can easily make changes including moving whole paragraphs around.

Start or Opening. How will you begin your story? What will the opening line be?

End or Conclusion. How will your story end? If your story or composition is intended to put forward an argument, what conclusions might be drawn?

You don't have to use each heading and you may think of some other headings that may be appropriate for a particular piece of work. Whatever headings you choose, you don't need to write huge amounts. The idea is the get the overall plan clear in your mind.

Composition

Once you have planned your composition, you can get started. The obvious place to begin writing is at the beginning, but if you're using a word processor, you can start anywhere you like. Whenever I'm writing an article, I almost always start in the middle. Why? Simply because I find the beginning the hardest. So, I start writing the middle bit first, then I usually go on to the ending. Often the beginning is the last part I write.

If you were handwriting a composition you are almost bound to begin at the beginning and finish at the end. But the beauty of using a word processor is that you can move about, writing the bits you want to write, when you want to write them.

You can click the mouse button anywhere in your composition and begin writing there.

Spelling

Most word processors have a spelling checker but you should be aware of its limitations.

First, if it highlights a word it doesn't necessarily mean that the word is misspelt. It simply means that the computer does not have that word in its dictionary. Names of people and places are unlikely to be in the dictionary so you will need to check elsewhere that it is spelt correctly.

Second, just because the spelling checker says all the words are spelt correctly, it doesn't necessarily mean that everything is correct. Consider this sentence, "*The book is over their.*" Each word is spelt correctly and a spelling checker would confirm the fact. But, the last word of the sentence is wrong. It should be, "*The book is over there.*" It's important to check your work very carefully as the computer will not always pick up errors like this.

If the Thesaurus chooses a word for you, make sure you know what it means.

Thesaurus

If you can't think of the word you want to use, type in something similar and choose the computer's Thesaurus to suggest something better. There's more information about a Thesaurus on page 111.

Letter writing

Letter writing seems to be a dying art – a fact that email has doubtless played a part in. Even though electronic communication is becoming more widely used and accepted, you will probably still have to write traditional letters on paper.

When writing a letter, it is important to get it right as it can say a great deal about you, regardless of what you actually write.

Your address. Normally your address will go at the top right of the letter, although some people now put it in the centre. Wherever you intend placing it, begin by typing it on the left, pressing Return after each line.

After entering your address, press Shift+Return a few times to ensure that only the address is on the right, not the whole letter.

(This applies to Word – with other, simpler word processors, you may have to drag the left Margin marker back to its original location before typing the body of the letter.)

1 Select your address by clicking the mouse at the top left and dragging to the bottom right.

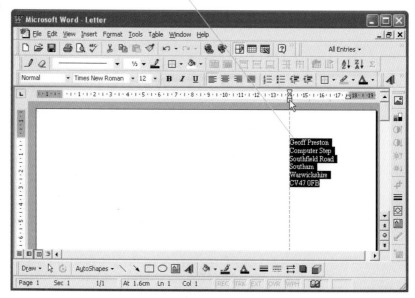

2 You could click the Right Align button to move your address to the right, but an alternative would be to go to the ruler just above the page on which you are working and drag the left margin marker to the right. Never move the address to the right using the Spacebar.

Recipient's address. The next part of the letter is the name and address of the person to whom you are sending the letter – the recipient. Leave a space after your address before typing the recipient's name and address which will remain on the left.

Date. Leave a space after the recipient's address and enter the date. Do not use abbreviations like 3/9/03 or 3 Sept 2003 as this gives the impression of laziness. 3rd September 2003 looks so much better and also avoids any confusion. In some parts of the world it is common practice to put the month before the day. So is 02/03/03 the 2nd of March or 3rd of February?

To make it look even better, you could superscript the letters *st*, *nd*, *rd* or *th* that come after the number.

Do not use a fancy font – stick to something clear and simple like Times New Roman or Arial.

1   Carefully mark the letters that come immediately after the number by clicking the mouse on the left of the letters and dragging to the right.

2   Click on the Format menu and choose Character... to open the Character dialog.

3   Click Super in the Position area and then click OK.

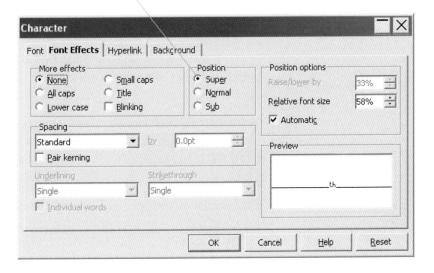

If this is a formal letter to someone you haven't met, refer to them as Mr/Mrs/Miss or Ms. Never call them by their first name.

Salutation. This begins with Dear. If you know the name of the person to whom you are writing, then use it. If not, it's Dear Sir, Dear Madam, Dear Sirs or Dear Sir or Madam.

Reference Number. If you have a reference number, (e.g. if you're writing to your bank), you should give your bank account number, if you're replying to a letter it may have a reference number to quote.

Now you can begin writing the body of your letter. Do not indent paragraphs: when you want to create a new paragraph simply press Return twice at the end of the last sentence of the current paragraph.

Closing. If you began your letter with the person's name e.g. Dear Mr Preston, then you should end with Yours sincerely. Note there is a capital 'Y' in 'Yours', but 'sincerely' does not have a capital letter. If your letter began with Dear Sir, Dear Sir or Madam etc., then finish off with Yours faithfully. Again, a capital 'Y' but no capital 'f'.

Although there are some variations of formal letter layout, you're generally better off sticking to these guidelines.

'Yours truly' is also a recognised closing, but it's hardly ever used now and sometimes the word 'Yours' is dropped altogether. With those exceptions, there are no other alternatives for a business letter. Especially not 'Regards from' or 'Lots of Love'.

Your name. Press return about 6 times and then type your name. Never give yourself a title like Mr Preston. If you need to let the reader know your gender, add the title at the end in brackets e.g. G Preston (Mr). On the line immediately after, you might want to clarify your position e.g. Student, North London College.

The large gap between the closing and your name is so that you can sign your letter using blue or black ink or ball point pen (or roller ball). No other colour is acceptable and it's generally not a good idea to use any other type of pen, especially not a fibre-tip marker.

Enclosures. If you are sending other documents with your letter, list them at the very foot of the page under the heading Enclosures or Enc:.

Letter template

It is likely that you'll write lots of letters for job interviews and college placements, so it's worthwhile creating a template.

Many word processors support special versions of their files called templates. Microsoft Word, for example, holds a range of templates in the Templates folder in Program Files/Microsoft Office.

When you have created your template, choose Save As... from the File menu and then select Document Template as the file type.

You could also create templates for faxes, memos and notes.

When you've finished your template, make it Read Only so you don't accidentally delete it.

You could create a new folder (e.g. Personal Stationery) and save your letter template in there.

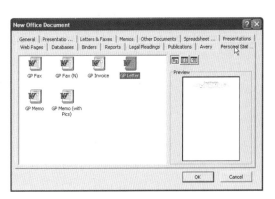

When you choose Start, New Office Document, you'll see a new tab. Click this to display your personal templates.

Websites

Authors

There are hundreds of websites dedicated to authors and their work. In most cases, entering the name of the author into the address pane of your browser will give you several sites e.g. entering 'Shakespeare' lists many sites including ShakespeareWeb which features the entire works of the Bard which you can search through. ShakespeareWeb is at `www.shakespeare.com/`

Writing Letters

For more information about letter writing, there are a few good sites including Write Letters at `www.writeletters.com/` which includes templates on a range of letter types including: Congratulations, Thank You, Complaint and Appointment.

Dictionaries

Microsoft's online dictionary at `http://dictionary.msn.co.uk/` provides spellings and definitions.

Save these sites in your Favorites folder so that you can access them quickly.

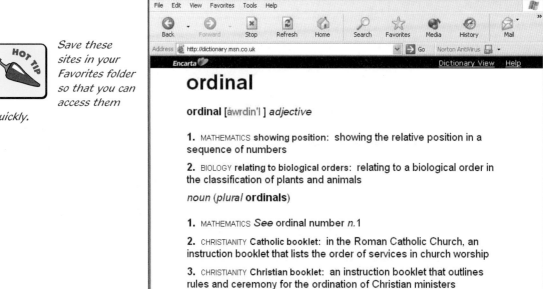

When you enter the site, type in a word and get all the information about it including part of speech and pronunciation.

A Thesaurus provides alternative words and phrases and the most famous example is Roget's Thesaurus. Visit `www.thesaurus.com/` for the online version of this valuable reference work. You can browse through the six main classes of words and search the main index. You can also select a dictionary from this site.

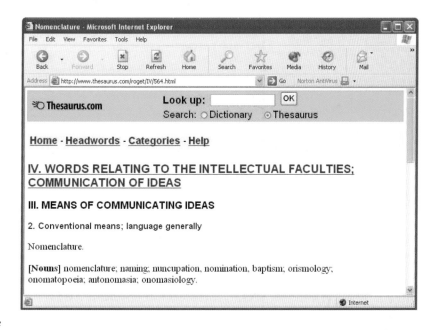

You could lose marks for incorrect grammar.

Grammar

The On Line English Grammar website at `www.edunet.com/english/grammar/` is a comprehensive guide to the correct use of English grammar. You can search for a part of speech, by a keyword, or look through the table of contents.

Alternatively, Grammar Bytes! at `www.chompchomp.com/` offers a range of interactive grammar exercises.

If you're not sure which book to read next, browse through the WHSmith catalogue.

Books

WHSmith's online bookstore at `www.whsmith.co.uk/` lists a huge range of books, many of which have one or more reviews written by readers. If you've read a good book, you can try writing a review which may be published online.

No homework?

Reading is probably one of the most important things you can do if you haven't got English homework. But why don't you write a review of some of the books you have read?

If you're going to try this, and you're likely to do it several times, it's worth creating a template.

The headings shown here are for guidance only. You might want to choose others and/or discard some.

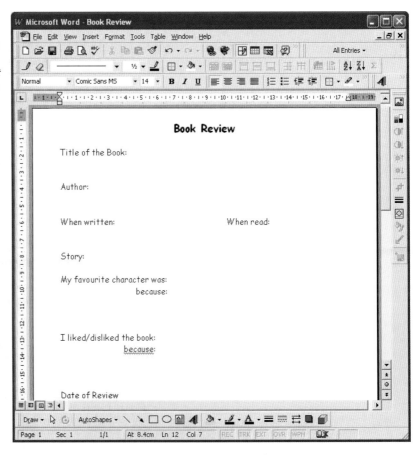

With a template like this you could either add your notes on-screen, or print out the form and complete it by hand.

If you think you might review lots of books, you could create a database of the books you've read. The headings shown here would be the fieldnames of your database. See pages 23–24 for more information about creating a database.

Food Technology

Food Technology is not just about cooking but includes nutrition and diet.

Covers

Chapter Thirteen

Menu

If you are creating a meal you may be asked to design a menu for it. You could do this with a DTP application and output your menu on something other than a flat piece of paper.

If you visit www.vikingdirect.com/ or www.vistapapers.com/ you will find pre-printed stationery in a variety of designs. These are really intended for small companies to quickly create their own corporate image but they are excellent for some of your school work especially if you don't have a colour printer. All you have to do is put black text in the right place to create professional looking results. Some suppliers even offer a sample pack containing one each of several designs.

Presentation is often as important as the actual content of your work.

These pictures are ©PaperDirect and are used with permission

Some word processors contain purpose-made templates for specific paper designs.

Even if you don't use special paper, try printing your menu as a two-fold brochure. Most document processors like GST's Publisher can support such a document but if you don't have a program that supports this format, many word processors can. (GST's website is www.gstsoft.com/.)

1 Create a landscape document with two pages.

2 Set it up in three columns with a gutter of about 20 mm between columns.

A simple plain sheet is quite effective, especially if you decorate it with a few lines or perhaps some pictures of food.

You'll need to think carefully about the order of the pages. On one side you'll have pages 1, 2 and 3 in that order, but on the other side you'll have page 4, the back of the folded brochure and the front cover in that order.

You can use large letter sizes and choose alternative font styles.

As the text automatically flows from one column to the next, you'll have to begin with column 1 on the first page, which will eventually be on the inside, then finish with the cover which will be on the right of the second page.

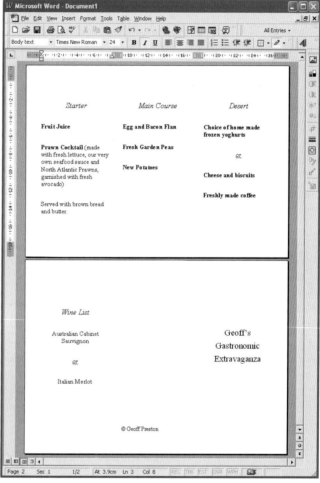

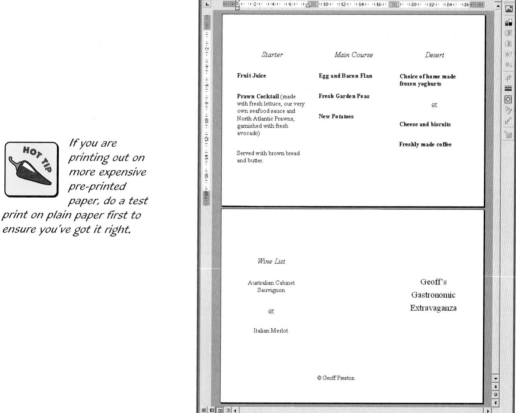

If you are printing out on more expensive pre-printed paper, do a test print on plain paper first to ensure you've got it right.

When you've printed your menu, place it on a flat surface so that the inside pages are facing up. Fold the right side so that the fold is between columns two and three. Now fold the left side over the top so that the fold comes between the first two columns. You should now see the front cover on top. If you turn the folded brochure over, you should see the back of the menu.

Recipe

A neat way of creating a recipe is to use a spreadsheet. The advantage is that all quantities will line up neatly and if you wish you can easily scale the quantities up or down to suit the number of people the meal is for.

1   In column A enter the quantity. Align these figures on the right.

2   In column B enter the unit (grams, spoonfuls, ml etc.) and align them on the left.

Some spreadsheet programs allow you to store several worksheets in one document. This feature will enable you to store several recipes in a single file.

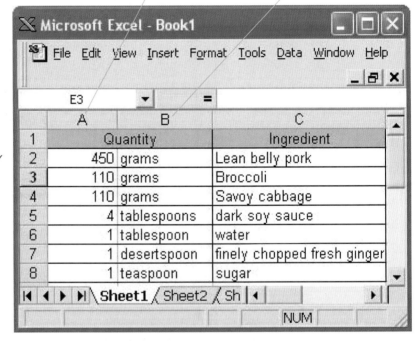

	A	B	C
1	Quantity		Ingredient
2	450	grams	Lean belly pork
3	110	grams	Broccoli
4	110	grams	Savoy cabbage
5	4	tablespoons	dark soy sauce
6	1	tablespoon	water
7	1	desertspoon	finely chopped fresh ginger
8	1	teaspoon	sugar

3   Put the name of the ingredient into column C.

Note that the cell borders are not drawn between column A and column B as you do not want to separate these items.

This recipe has been written into a spreadsheet in its simplest form, but you can improve upon this idea.

If you want to be able to quickly alter the quantities so that the recipe is for a different number of people:

In a spare cell enter the number of people the recipe is for. (I've used cell E2.)

	A	B	C		E	
1	Quantity		Ingredient		Serves	
2	900	grams	Lean belly pork		4	
3	220	grams	Broccoli			
4	220	grams	Savoy cabbage			
5	8	tablespoon	dark soy sauce			
6	2	tablespoon	water			
7	2	desertspoon	finely chopped fresh ginger			
8	2	teaspoons	sugar			

Microsoft Excel - Book1

File Edit View Insert Format Tools Data Window Help

A2 = =225*E2

Sheet1 / Sheet2 / Sheet3

Ready / NUM

Spend time making sure you enter the correct quantities per person.

2 Instead of entering the actual quantity in column A, enter the quantity for 1 person and multiply it by the cell containing the number of people the recipe is for.

For example, the recipe shown opposite was originally for 2 people. Instead of entering 450 grams of pork, enter 225 grams (the amount for 1 person) and multiply that by the cell containing the number of people – in this case E2. So, in cell A2 would go the formula:

=225*E2

Do this in all cells in column A.

Now, by simply changing the number of people in E2, the whole recipe will recalculate the quantities for that number of people.

Websites

Apart from the cookery/recipe sites, of which there are hundreds, the Web has several Government food sites which deal with nutritional and food safety information.

British Nutrition Foundation

This fun site at www.nutrition.org.uk/ features polls, advice and features of general interest. When you get to the BNF website, click on the 'Pupil Centre' link to display a range of links divided into 3 age groups covering diet, health and nutritional advice.

If you enter the word 'Food' in the address bar or search engine you will probably get sites which deal mainly with recipes.

BBC – Recipes

There have been cookery programmes on TV for as long as I can remember. This site at www.bbc.co.uk/food/ features recipes from popular programmes, food information and articles about your favourite chef. You can even sign up for a regular newsletter.

Food Standards Agency

This Government agency at www.foodstandards.gov.uk/ deals with food safety and standards. There are regular newsletters as well as valuable information on food additives.

No homework?

Nutrition Database

If you look on the packets of any food (that includes sweets, cereal and crisps) you'll find nutritional information in the form of a table. In fact there's usually two tables, one for the nutritional content of 100 grams of the food and one for the values for a typical serving.

You could build a database of additives so that if you see an additive in an ingredients list, you can look it up to find out what it is and what it's used for.

The fields you are likely to need for your database are:

Energy. This value is usually given in kJoules and kCalories. Decide which one you want to use and stick to it. Don't enter some foods in kJoules and others in kCalories.

Protein. This is usually expressed in grams. When entering the value, just enter the number and not the letter 'g' or 'grams' as well.

Carbohydrates. There are often three values given here: the total carbohydrate content and the figures for sugar and for starch. Decide which you want and stick to it.

Fat. This heading frequently has two values: the total fat and the saturated fat. Sometimes it also lists the unsaturated fat which is the total fat minus the saturated fat.

Fibre. Otherwise known as roughage. This is an important field to include and the value is expressed in grams.

There may also be a list of vitamins and minerals which may be included in the database if you wish, but be aware that if you add too many fields you'll require an enormous amount of data to fill them.

When compiling the nutritional database, enter a little at a time or you will soon become bored with it.

Finally, you'll need to include the name of the food so that you'll know what each record relates to.

As with all databases, begin by constructing the master page. This is where you lay out the fields and enter all the details that are common to all records. In particular this will be the fieldnames, headings and any units that you want displayed.

Some database programs allow you to enter the data in either a table or individual records.

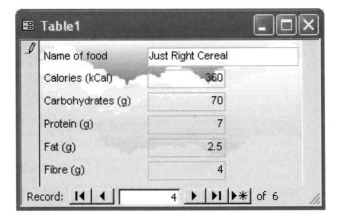

You can use this database to calculate your daily intake.

The record above was created using Access, which is part of Microsoft Office. Once you've created the fields in a table, you need to declare the type of data that will be entered in each field. The name of the food clearly has to be text. The Calories field will only contain whole numbers so you can set that data type as Integer (which is a whole number without any fractions). All the other fields used here could contain fractions of a number and so they should be declared as being a Real number – that is a number that can contain decimal places.

Once the fields have been set up, you can enter the data either in the table or in the individual records.

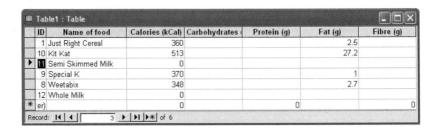

The advantage of displaying the data in this format is that it's easier to make comparisons.

Geography

Geography relies heavily on graphics – in particular, maps, charts and graphs.

Covers

Chapter Fourteen

Maps

In order to draw on a computer, you clearly need a drawing program. Paint, which is supplied with Windows, can be used for some simple drawing but, as the name implies, it's really a painting program.

Coastline Maps

You may be asked to draw a map of any area from a whole continent down to a few square kilometres. Creating a map of a coastline is easy using the following method.

If you have a scanner, scan the map you want to use as your starting point. The scanned image can be placed in the background of a drawing program like GST's Draw, traced around and then deleted leaving the outline drawing. (For more information about Draw, visit the GST website at `www.gstsoft.com/`.)

If you haven't got a scanner, trace the coastline on the thinnest tracing paper you can find and stick it to the front of the monitor. You should be able to see through it enough to be able to follow the line around.

After scanning the image and saving the picture:

1 Run your drawing program and open a new document.

2 Choose Import File from the File menu.

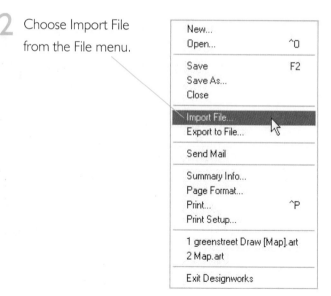

3 Import your scanned map.

4 Zoom the drawing so you can see the outline of the map clearly.

Ideally the coastline should be one continuous line with the last point plotted over the start point.

5 Go to the View menu and choose Grid. Ensure the Snap to Grid feature is off (unticked).

6 Choose the Freehand Pencil tool and begin carefully drawing around the coastline of the map.

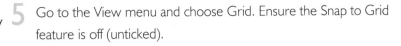

If your hand begins to get tired, have a rest but try to restart where you stopped.

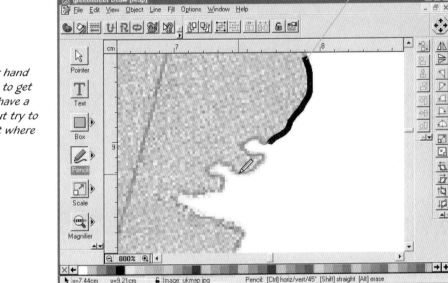

The coastline may look too thick on the map opposite, but it is scaled to 800%.

When you've finished you can then go around it making any minor adjustments. You will see that when you have finished, the line will be drawn to the thickness you specified but will have lots of dots on the line. These points are called 'nodes' – you can drag them around by moving the mouse pointer exactly onto one of them, clicking and holding the left mouse button and sliding it into a new position.

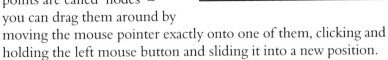

Finally, go round the coastline drawing in any of the islands.

When you have completed the outline and added the islands, you can then delete the original scanned map leaving you with just the outline of the landmass and any nearby islands it may have.

This method also works well for county or country borders.

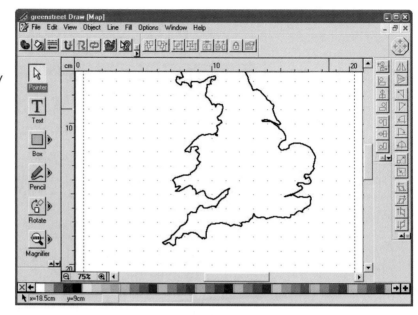

The resulting outline map can then be detailed. Exactly how you do this will depend on precisely what task you've been given. You may need to add county or country boundaries, towns and cities, landmarks such as roads and buildings etc. This type of map is called a Political Map as it shows the bits we (humans) have added.

This could also be the start of a weather map.

Alternatively, you may need to make it into a Physical Map which shows all the natural features such as rivers and mountains.

You may wish to insert your finished map in a text editor like WordPad. If you do, save it in the usual way and then export it as a Windows Metafile. This is because WordPad may not be able to read files saved in the format your drawing program uses, but it can read Windows Metafiles.

Street maps

Street maps are surprisingly difficult to draw really well, but here's a method that's just about foolproof.

Using Draw or a similar vector graphics program:

1 From the View menu, choose Grid.

2 Choose a fairly coarse grid, (in this case I've used .5mm divisions) and, most important, check the box labelled Snap to Grid.

3 Next, increase the line width. Normally you would draw with a line about .5mm, but here, I've used a 4mm think line.

Whenever you begin a drawing, spend a few moments setting up the various options. It will save you a great deal of time in the long run.

Once you have done this, you're ready to draw. Draw the roads with a single, thick black line. You'll notice that the grid will force you to draw at the grid points and if you try to draw somewhere else, the pointer will jump to the nearest dot on the grid.

Place intersections carefully so that 'T' junctions meet to form a 'T' and not something closer to an 'X'.

You will need to use a combination of straight and curved lines to complete the map.

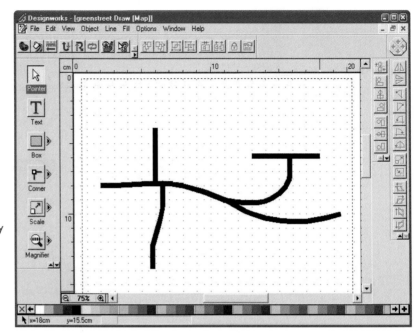

If you later find you've made a mistake you may have to delete everything back to this stage.

Spend time on this part. Get the roads positioned as accurately as possible. When you've got everything as you want it, choose the Editing tool (the arrow at the top) and drag a rectangle over the entire map to select it. (Another way is to choose Select All from the Edit menu.) Then:

1 Copy (Ctrl+C or choose Copy from the Edit menu).

2 Paste (Ctrl+V or choose Paste from the Edit menu).

What you have done is copied black lines on top of existing black lines. You won't see any immediate difference at this stage, but do not deselect any part of the drawing yet.

3 Change the line thickness to something less than you previously had. If you began with 4mm, then 2mm or 3mm would be about right.

4 Change the line colour to white.

5 Select All again, and then Group all of the elements so they are 'glued' together as one object. This will prevent you from accidentally moving or deleting one of the lines. (Ctrl+G is the Group Objects shortcut used by most programs of this type.)

This often doesn't look very good on-screen, but when it's printed it looks great!

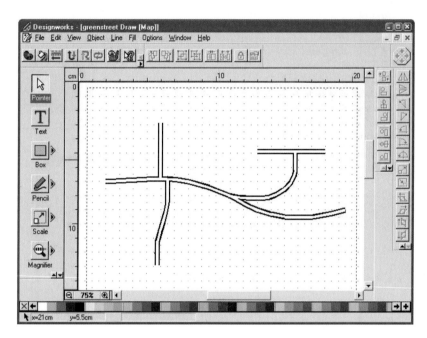

Now you've got roads, edged in black, ready to be detailed.

You can now add the road names, and any geographical symbols necessary.

Once you are used to this process you can take it a stage further. Main roads are usually shown wider than minor roads and are frequently shown in a different colour. If you want to try this, begin drawing the main roads in, say, 6mm black. Then the minor roads in 4mm black. Next, select just the minor roads. Copy, paste and change colour and thickness. Finally select the black lines of the major roads. When you've copied, pasted and changed the colour and thickness you'll have a very good representation of a geographical area.

Diagrams

You will frequently be asked to produce a diagram or a series of diagrams to demonstrate how a particular process works. In many cases Paint can be used to draw the diagrams.

1 Begin by laying down some guidelines.

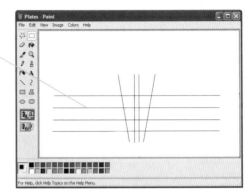

If you have to draw several similar diagrams, use Paint as you can simply draw the first, save it, alter it to make it into the second, save it, alter it for the third and so on.

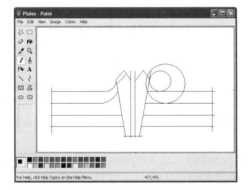

2 Add more detail to the guidelines. Creating curves is often a case of drawing a full circle and rubbing out the bits you don't want.

If the left side is simply a mirror of the right, you can work on just one side and then make a copy, flip it and match the two parts together.

3 Adding the colour is where the fun begins. You can also start some of the details like, in this case, the arrows to show the direction of the forces.

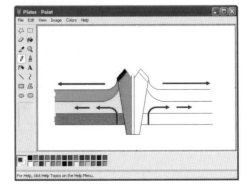

When adding labels to a diagram, draw a single, thin, black line from the label to the part.

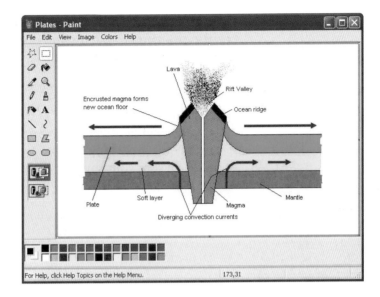

4. Add the final details and label the major parts. It's now ready to be either printed or included in a word processed account of the forces that cause Earthquakes.

Having completed one diagram, it's often a fairly simple task to alter it to make it into a similar, but different diagram.

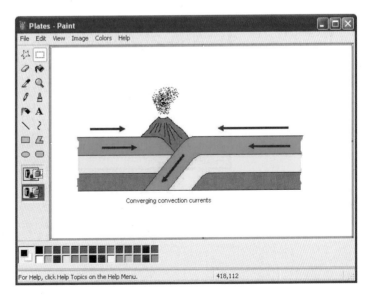

Graphs

Graphs are frequently used in Geography to show anything from population figures to the use of exhaustible reserves.

Creating a graph by hand can be a lengthy process, but if you've got a computer program to do it for you, it could be done in minutes.

Graphs require some data to be entered and the best program for this is either a dedicated graphing program (i.e. a program specifically for producing graphs) or a spreadsheet, but before you begin, make sure your spreadsheet can output graphs.

Data is normally entered in a vertical column. Labels can be inserted at the top of each column and alongside.

Graphs are also used in Science and Mathematics.

1 Start your spreadsheet or graphing program.

2 Enter the numbers you want for your graph in Column B

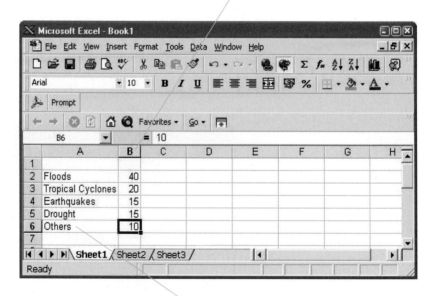

3 In Column A, enter any labels you want to include on the graph.

Choose the Create Graph button which, in the case of Microsoft Excel, is on the Button bar at the top of the window.

This opens a wizard:

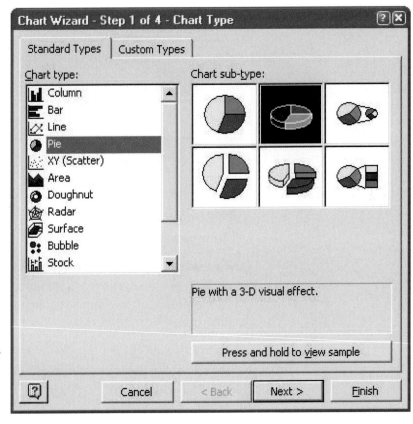

Always include a title and labels explaining what the data represents.

The first task is to choose the type of graph or chart you want to use. As a general rule, if the numbers you've entered add up to 100, then a pie chart is frequently the best option.

After selecting the appropriate graph type, Excel lets you make other choices like showing the chart in 3D or separating one or more of the segments. When you've made your choices, click the Next button.

Continue through the wizard adding titles and labels where necessary.

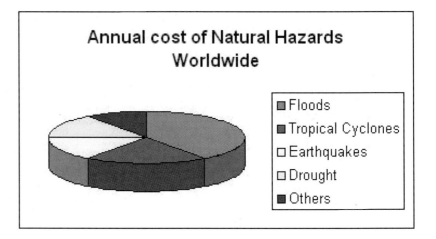

The final graph can be printed or saved as a picture and inserted into a document.

If you keep the graph in the spreadsheet, any alterations to the data will automatically change the graph.

Other ways to create a graph

Some word processors have graphing facilities built in. If, for example, your graph is going to be included in a Microsoft Word document (to illustrate an essay), you can create your graph in Word.

In your Word document, open the Insert menu on the menu bar and choose Object. A dialog opens listing all the possible objects you could insert. Choose Microsoft Graph.

This will open a window very similar to a spreadsheet. Enter your data into the grid and choose the graph type from the Graph button on the Button bar. The graph will be drawn at the position of the cursor and can be aligned left, right or centre and have text wrapped around it just as you can with a picture. (See page 54.)

To edit the data, simply double-click on the graph and it will open the grid showing the original data entered. Altering the data will automatically alter the graph.

Websites

Geography students are spoilt for choice on the Web. There is a huge amount of material including maps, flags, symbols and articles about all aspects of Geography.

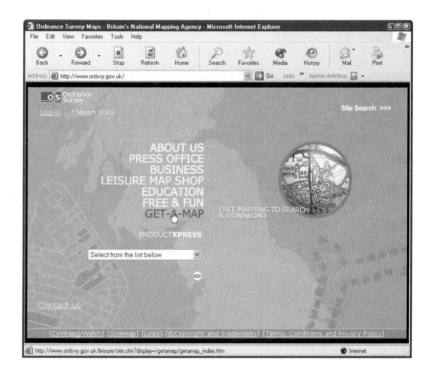

Ordnance Survey is Britain's National Mapping Agency (`www.ordsvy.gov.uk/`) and has some excellent resources for students of all ages.

If you want to find ready-made maps, try visiting `http://geography.about.com/` which is one of the sites from the enormous About.com network. Although aimed primarily at US users, there is a great deal of material of interest to UK schools.

The National Geographic is a long established periodical for geographers. Their website is `www.nationalgeographic.com/`.

Finally, Weather is an important part of geography. Typing in 'weather' into the browser's address bar will list dozens of sites providing forecasts and information about weather systems.

Entering 'geography', 'cartography' or 'maps' into the address bar will also list dozens of useful sites.

No homework?

Returning to the theme of graphs, here's an interesting long-term project. Find a parallel-sided beaker and a thermometer. It is important that the beaker has parallel sides all the way up. Stand the beaker in the garden where it can't be knocked over and hang the thermometer nearby.

Empty the beaker after you've measured the depth of water.

Each evening, when you return from school, measure the depth of water and take a note of the temperature.

It would be better if you collected data for a month, or even a year.

The figures can be placed into a graphing program and after a week you'll have enough information to draw a graph.

	Rainfall	Temperature
Sun	0.5	2
Mon	0	3
Tue	1	6
Wed	1.5	5
Thu	0.5	4
Fri	0	7
Sat	0	8

If you want to type the degree symbol (°), ensure Num Lock is on, press and hold the Alt key (not Alt Gr) and type 0176 on the Numeric keypad.

Some programs that can produce graphs offer a huge selection of graph types. If you use Microsoft Excel, after entering the data, click the Graph button and choose Custom graphs. Scroll down the list and choose *Line - Column on two axes*. This will enable you to display the rainfall as bars and the temperature as a line on the same graph.

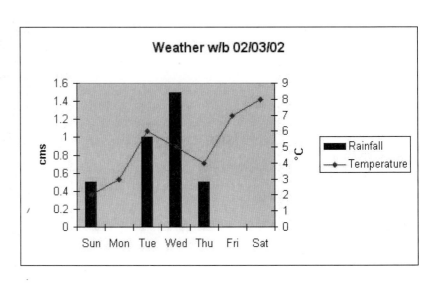

History

You don't have to write your essays on single lines which span the entire width of the page. Using columns is an interesting alternative.

Covers

Chapter Fifteen

History News

Writing an essay will be one of the most common homeworks for History so you'll probably be using a word processor to complete most of the work. But rather than simply typing lines of text that run from the left margin to the right, why not use columns? Most good word processors can support multi-column text, and some even supply a template to get you started quickly.

Most people will have an A4 printer and will therefore be printing on A4 paper – 297mm tall by 210mm wide. This paper size can easily take two columns, and provided you don't use large text, you can fit in three.

In Microsoft Word, for example, open the Format menu and choose Columns to open the dialog in which you can set your preferences.

Columns don't have to be the same width.

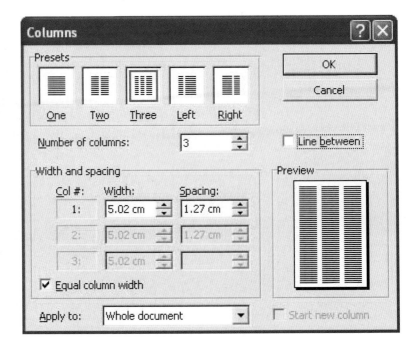

Don't choose more than 3 columns on A4 paper.

Choose the number of columns and decide on the other options (like whether you want a line to separate the columns) and click the OK button. The resulting document will be set up to automatically flow the text from one column to the next. In other words, begin typing in the left column and when you get to the bottom, the text will automatically start to be placed at the top of the next column.

The next step from writing in multiple columns is to write your essay like a newspaper article.

Most good word processors can work with columns. Alternatively use a DTP program.

If you're using three columns, don't use text larger than 10pt as you will find you'll only get a couple of words across a column width.

History News

Websites

The Internet offers a huge number of good quality history sites for your research. Some are general sites whilst other are about one particular topic or time period.

Typing 'History' into the search panel or into the address bar provides a range of general History sites. There is a large index of resources at www.ukans.edu/history/VL/. On the title page is a list of topics including Architecture, Art, History of Cartography, Religious Missions, Film History, Holocaust, Indigenous Peoples, Islam, Labour and Business History, Maritime History, Military History, History of Science, Slavery, Urban History and Women's History.

The History Channel

The History Channel is a popular television channel featuring documentaries from all historical periods. The website at www.thehistorychannel.co.uk/index.htm not only features support material for programmes but also a search engine that links to other sites.

Try searching for a topic like 'French Revolution'.

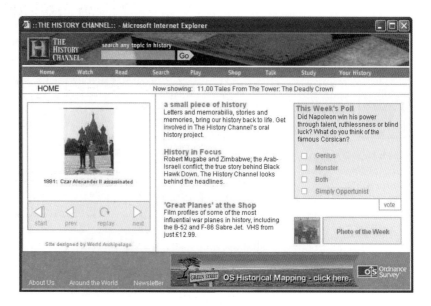

At the top of the window is a panel into which you can enter any topic or period in history and a search will be made for all the articles available.

A search for Marie Antoinette, for example, results in 13 documents being found which relate to her and life in France during that period. Click on one of the links to take

you to the article. You'll find that there may be several articles on the same subject. If this is the case, you may have to look at more than one before you find the exact information you want.

When entering names into a search engine, ensure the spelling is correct.

Britannia

Britannia is a huge resource with an index and a search engine.

If you scroll down the first page you will find dozens of historical periods and characters which, when clicked upon, will take you to a quality account of the period with excellent pictures to accompany the text.

You can also find the subject you're researching by typing a period or character into the search engine. Visit Britannia at `http://britannia.com/history/`.

No homework?

It can be quite difficult to put different world events into context. If you learn about events, say in Germany during the 1914–1918 period, one can lose sight of other important events that happened during that period. For example, the Russian Revolution of 1917 started whilst the rest of the world was at war.

There are a couple of ways to create a timeline, but one method that works quite well is using Paint which is supplied with Windows. When using this program you'll need to take care to get it right first time as alterations can be a little tricky .

As a timeline is quite long, and very narrow, it's best to draw it vertically as you would normally complete your written account on paper with the longest side vertical (portrait).

A vertical timeline also means you don't have to rotate the text.

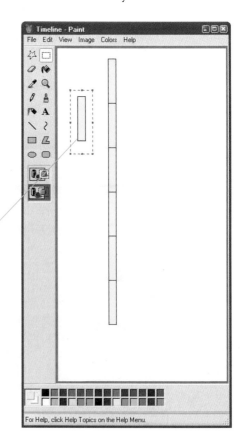

1 Begin by drawing a thin rectangle or line to represent a number of years. (In the example, I chose to represent 10 years with a small rectangle.)

2 Choose the Select tool and select the rectangle you've drawn.

3 Make a copy of it and drag it into position. Continue copying and pasting so that you have several rectangles which span the years you want to cover.

4 Next, add the main dates on the left of the timeline and then the events on the right.

If you have an event that covers several years, you may need to bracket the years in the way shown here for the two World Wars.

Several events in consecutive years may cause a little overcrowding and you may need to add lines or arrows to point to the exact year as shown here for the Russian Revolution. Too much overcrowding might mean that you have chosen the wrong scale. In other words you need to use longer rectangles for shorter periods of time.

You could make the bar wider so that the dates could go on the line leaving space to put events on either side.

Alternatives

This timeline works well for all periods of time from millions of years to just a few hours, providing you begin by choosing the correct size line for a particular time period.

If you have a spreadsheet program you can use that to create a timeline. Again, a vertical line is generally easier, so begin by making column B quite narrow (say 5mm) and colour it to make the line. Dates can then be places in column A and events in column C, which should be made wider. Using a spreadsheet, it will be much easier to edit your timeline.

This is shown as an example, but you would normally create a timeline for just two or three topics or events.

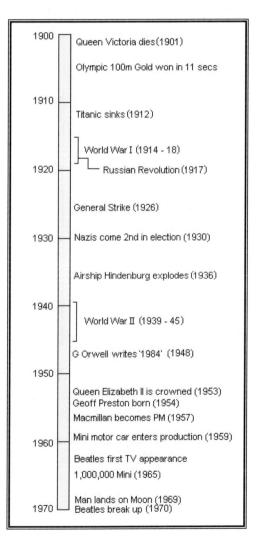

1900	Queen Victoria dies (1901)
	Olympic 100m Gold won in 11 secs
1910	Titanic sinks (1912)
	World War I (1914 - 18)
1920	Russian Revolution (1917)
	General Strike (1926)
1930	Nazis come 2nd in election (1930)
	Airship Hindenburg explodes (1936)
1940	World War II (1939 - 45)
	G Orwell writes '1984' (1948)
1950	Queen Elizabeth II is crowned (1953)
	Geoff Preston born (1954)
	Macmillan becomes PM (1957)
	Mini motor car enters production (1959)
1960	Beatles first TV appearance
	1,000,000 Mini (1965)
	Man lands on Moon (1969)
1970	Beatles break up (1970)

Census data

If you're studying local history then you could extend your work by downloading some census data for your local area. Every 10 years (the year ending in 1 e.g. 1981, 1991, 2001) the UK holds a census to establish exactly who is living in the country. After 100 years the data is released to the general public. Some of it can be downloaded from the Web. One site, `www.censusuk.co.uk/`, offers a limited amount of free census data.

My great grandmother's name was Dyke and so I entered that in the search panel together with her birthplace, and the following day received an email with a text file attached which can be opened using Notepad.

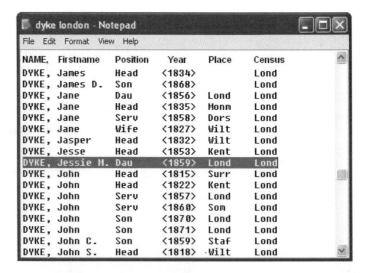

NAME,	Firstname	Position	Year	Place	Census
DYKE,	James	Head	<1834>		Lond
DYKE,	James D.	Son	<1868>		Lond
DYKE,	Jane	Dau	<1856>	Lond	Lond
DYKE,	Jane	Head	<1835>	Monm	Lond
DYKE,	Jane	Serv	<1858>	Dors	Lond
DYKE,	Jane	Wife	<1827>	Wilt	Lond
DYKE,	Jasper	Head	<1832>	Wilt	Lond
DYKE,	Jesse	Head	<1853>	Kent	Lond
DYKE,	Jessie M.	Dau	<1859>	Lond	Lond
DYKE,	John	Head	<1815>	Surr	Lond
DYKE,	John	Head	<1822>	Kent	Lond
DYKE,	John	Serv	<1857>	Lond	Lond
DYKE,	John	Serv	<1860>	Som	Lond
DYKE,	John	Son	<1870>	Lond	Lond
DYKE,	John	Son	<1871>	Lond	Lond
DYKE,	John C.	Son	<1859>	Staf	Lond
DYKE,	John S.	Head	<1818>	·Wilt	Lond

The file is in columns with headings at the very top. In this case, surname, first name, position in the house, year of birth, birthplace and location at the time of the census.

The data isn't a great deal of use in this format, but it is ready to transfer into almost any database program so that you can carry out your own database searches at home.

Before doing so, it's worth amending some of the data. Put the cursor at the top of the text and open the Find and Replace dialog which is found in the Edit menu.

Because the width of a column in the text file is limited, there have been some abbreviations made which could now be expanded (e.g. 'Lodger' has been reduced to 'Lodg').

If you originally searched for a name (e.g. Dyke) then every record will contain that name. Remove them from the textfile first by searching for 'DYKE,' and replacing with nothing.

1 Enter the word you want to find.

2 Enter the word you want it replaced with.

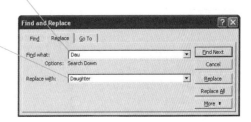

Save the amended file with a different name so that you still have the original.

3 Click the Replace All button and all occurrences of the word will be replaced by the new word.

Once you've got the data in the format you want:

1 Create a new datafile and place fields in the order shown at the very top of the census text file. The surname (which in this file is always DYKE) and the census place (which in this file is always London) have already been removed from the original datafile.

2 Import the data by choosing Import from the File menu on the database menu bar.

I've used the fieldnames that were originally used in 1881 even though some of those fields would not be acceptable today.

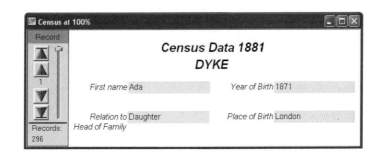

Once you've got the data into a database, you can perform searches (like listing all the people born between two dates) or make comparisons between data from more than one census.

Registered members of this site can search for data for a particular area. Entering the name of a street (in this case Prospect Place in Tottenham, North London) will result in a list of residents. This data is from the 1881 census.

If you're going to search this Internet site for the residents of a particular place, be very careful with the spelling. 'Street' and 'St' would be regarded as two different places. Spellings can also vary – there were 6 ways of spelling Birmingham in 1881!

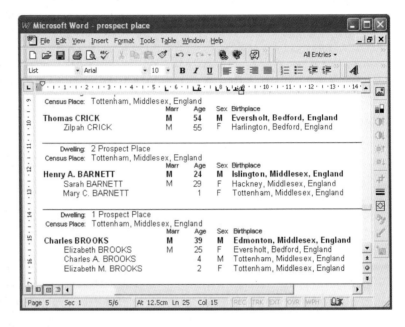

If you live in an old house you might be able find out who was living there years ago.

This data contains many more fields. Begin by inserting the fields into the blank database (in the same order as those in the textfile) before importing the data.

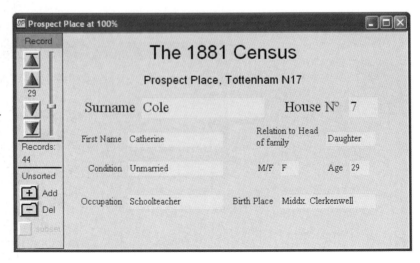

Modern foreign languages

If you study a foreign language like French, German, Italian or Spanish and want to use your computer to complete your homework then you'll need to learn how to enter the special characters used in those languages.

Covers

Chapter Sixteen

Special characters

Anyone who has studied a foreign language will soon notice that many characters (letters) seem rather strange. Some languages, like Japanese, contain completely different characters which bear no relation to anything we're used to reading. Some, like Russian, contain many letters that look vaguely similar to ours. Most European languages use the same letters we use, although some appear to have little bits added to them.

French is a good example. There are several characters, mainly vowels, that have 'dashes' over the top. These are called graves (è) and acutes (é). Sometimes, some letters appear to be wearing a hat called a circumflex (ê).

Some German characters have two dots above them known as umlauts (ë). There is also the character Germans use for a double-S called a sharp-S (ß).

These characters, and other special symbols like the degree sign (°) and the copyright symbol (©), do not appear on the keyboard and so we have to find another way of entering them.

Of the three different ways of entering special characters, Character Map, which is supplied with Windows, is probably the easiest method for beginners.

If you are going to use this program frequently, create a shortcut on the desktop.

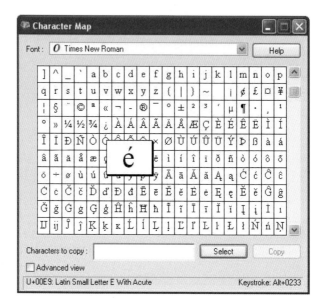

The Character Map program can be found by clicking on the Start button, then All Programs, going to Accessories, System Tools and double-clicking on Character Map.

It appears on the desktop as a grid and the first thing you should do is choose the font style you're currently using. Then:

1 Choose the character you want to use by moving the mouse pointer onto it and clicking the left button.

2 Click the Select button and then the Copy button.

3 Move to the document you are writing, position the cursor where you want the character to be placed and paste it using Ctrl+V.

You may wish to copy and paste several characters at one time in which case double-click on each character you want and they will appear in the panel at the bottom of the Character Map window. When you have all the characters you want in the panel, press the Copy button and go to Step 3.

It's sometimes quicker to enter the characters in English, and then, when you've finished and you're checking your work, change them to the correct foreign characters in one go.

Using the keyboard

An alternative way of entering these special characters is to enter their number. If you look at the bottom right of the above window, you will see a code. In the case of the lower case é, it's Alt+0233. To enter a character in this way, press the Num Lock key. Hold down Alt (not Alt Gr) and enter the code using the Numeric keypad on the right of the keyboard (not the numbers above the normal letters). Release Alt to insert the accent.

If you're going to use this method, then you've either got to remember each code or refer to a table. Here are the main characters you'll require for French:

Copy this chart neatly and pin it on your wall above your computer.

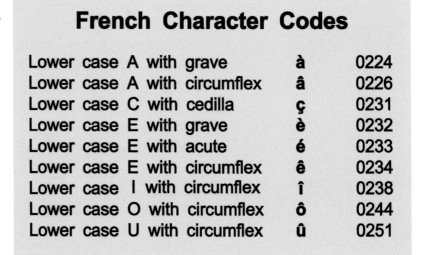

French Character Codes

Lower case A with grave	à	0224
Lower case A with circumflex	â	0226
Lower case C with cedilla	ç	0231
Lower case E with grave	è	0232
Lower case E with acute	é	0233
Lower case E with circumflex	ê	0234
Lower case I with circumflex	î	0238
Lower case O with circumflex	ô	0244
Lower case U with circumflex	û	0251

For this method to work, ensure Num Lock is on.

Spell-checking

Answering questions or writing essays can both be done using a word processor.

If you're answering questions, always include the question in the answer so that when you come to revise your work you'll know what it was about. A page containing a list of one-word answers won't help you.

If you have Windows, you will have WordPad which is supplied with most versions of Windows (including 98, Me and XP) and is adequate for most of this work. What WordPad doesn't include is a spelling checker. For writing in English, this may be considered a disadvantage. For word processing in a foreign language it may be better if you don't have a spelling checker as it could interpret every foreign word you enter as a potentially misspelt word.

This is because many word processors bought in the UK only contain a list of English words against which the software checks each word you enter.

A word processor sees a word as being a group of letters with a space on the left and a space or punctuation mark on the right.

Adding a dictionary

Some word processors are supplied with dictionaries for languages other than English. If your word processor only has English available, it may be possible to install another language.

Microsoft Word is capable of running several dictionaries that can not only spell-check your work but can check grammar too.

With a second dictionary installed:

1 Click on Tools on the menu bar and choose Language, Set Language.

2 Choose the language you want to use and, if the dictionary for that language is available, it will automatically use that dictionary for spell-checking. If not, it will revert to English.

If you want to install a foreign language dictionary, visit Alki at www.proofing.com/. They have dictionaries in over 20 languages which can be installed into Microsoft Word.

Websites

There are many websites which will help you when learning a foreign language. Some will provide translations and some include information about the country.

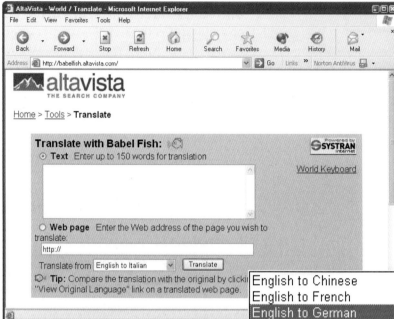

Don't use this site just to get your homework finished. You need to learn and understand.

One of the best translation services is from Babel Fish at AltaVista (`http://babelfish.altavista.com/`) which allows you to enter up to 150 words, in one of ten languages after which you can choose to have it translated into English, French, German, Spanish or Italian.

This is a quick way to get single words or short pieces of text translated, but it's not for translating all of your homework.

Find out about

Of the sites that tell you about the country, most are intended for would-be tourists. But nevertheless, many offer a fascinating insight into the habits of the natives.

Begin by opening the browser and entering the name of the country or the name of the language into the address bar to start a search. You'll get a list of some really interesting sites like this one for France at `www.francekeys.com/`.

There are lots of really good websites to help you with your foreign language studies.

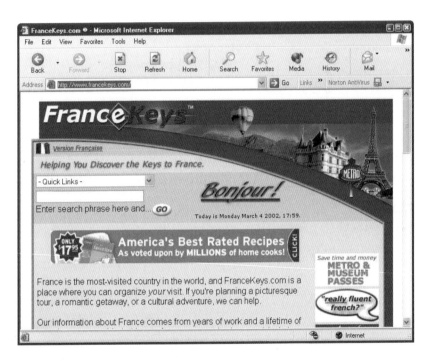

Some of the other sites worth visiting are:

All about Spain	`www.red2000.com/spain/`
In Italy	`www.initaly.com/`
Guide to Germany	`www.guide-to-germany.com/`
Lingolex	`www.lingolex.com/`
Languages for Travellers	`www.travlang.com/`
Zip Zap France	`www.zipzapfrance.com/`

No homework?

A foreign penpal may also want to improve their foreign language so don't be surprised if you receive replies in English!

A pen friend

Once you've mastered typing in a foreign language, you might like to find a penpal who you can communicate with in that language.

There are three main methods in which you could do this:

Email

If you have an email address you could send messages to someone in a different country. Your email address would be sent with the message so they would be able to reply. This is the best method for those who are not very fast on the keyboard.

Chatroom

You'll need to be quite fast entering foreign language characters if you want to have a live keyboard chat. If you do succeed in finding someone, it's usually best to have a private chat as you don't get lots of people butting in. Visit TeenChat at `www.teenchat.co.uk/`.

Instant Messaging

This is a little more sophisticated than chatrooms because you have a selection of contacts that you choose. When they go online a message pops up telling you, so that you can decide if you want a chat with them. Similarly, when you're online, they will know. One of the most famous instant messaging programs is ICQ (I seek you) which you can download for free from `www.icq.com/`.

Before you can practise your mastery of a foreign language you will need to find someone to message with.

Begin by checking with your teacher who may have a list of foreign students who you could write to, or a contact who could put you in touch with someone.

If you can't get the name of a penpal from school, you could try searching on the Internet, but proceed with caution and let your parents know what you intend to do. Begin by entering 'penpal' into the address bar of your browser. You'll find lots of sites offering the chance to find a penpal. One of the these sites, especially suitable for young people to find an international penpal, is Interpals which may be found at `www.interpals.net/`.

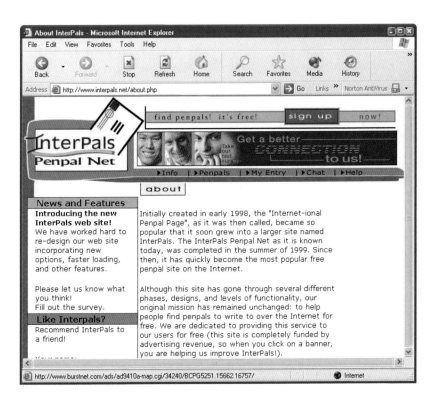

For safety's sake, do NOT give away your home address or arrange to meet strangers you have been chatting with on the Web.

You'll need to register but before you do, take time to read the introduction. Get your parents to read it too.

When you have registered you can find a penpal very easily by entering a search. Fields include age range or exact age, gender or country. When you choose a possible penpal, send them an introductory email.

Other good penpal websites include:

PenPals	www.penpals.com/
PenPALS	www.bestmall.com/penpals
Teen Penpal Classifieds	www.enn2.com/teenclass.htm
Techno Teen	www.technoteen.com/teen/
Penpal Network	www.penpal.net/

Build a dictionary database

A useful way to build up a vocabulary is to enter words you come across into a database. This is something that you can add to from time to time – perhaps when you get home on the day you have Modern Languages you could add some new words.

Don't try to enter all the words you know in one go because you'll get bored with it. Do a few at a time.

A database is a way of storing similar, ordered information. There are lots of database programs and they all work in a slightly different way although the principle is the same. Begin by creating a layout card. This is sometimes called a template. On the template you can insert a title that will appear on every page. You will also need to insert at least two fields – one for the English word and one for the foreign equivalent. If you are studying more than 1 language you could add further fields.

Each field would normally have a name – not surprisingly this is known as the fieldname.

Some databases even allow you to add sound samples. You could record your voice speaking each word so that when you click on the word it is played back to you.

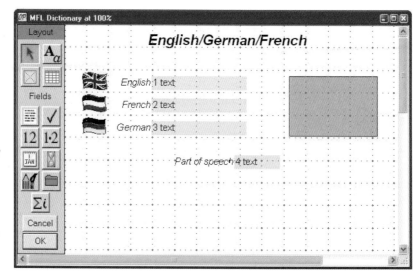

You could go further by adding a field for the part of speech, or the tense. It could be made more visually interesting by adding flags of the countries for each of the main fields. You could even leave space for a picture to represent the word. You can dress it up as much as you like.

Once you have completed the layout, you can then move on to entering the words in records. Each record contains information about one 'thing'. In this case, it's one word. When you've added the English word and the translation(s), enter the part of speech and then any picture and sounds you want to include.

Move on to the next record and do the same thing again for a different word.

If you don't have a database, you could do a similar exercise with a spreadsheet.

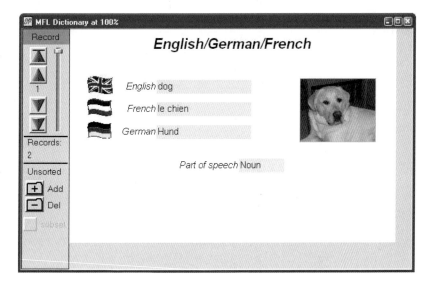

If you don't know the French or German translation for a particular word, leave it blank and come back to it later when you've looked it up. You might even discover some foreign words which you don't understand. You could enter these first and then insert the English word when you've found out what it is.

This is a fun way to build up a vocabulary of words. But it's more that just fun – it has real value. If you enter words as you come across them, by the time you get to the end of Year 9 you'll have quite a large dictionary which you can then use to look up words.

Searching your database

Once you've built up a decent database of translations, you can use the search facility provided in each database to look up a word.

Be consistent. If you decide the part of speech should be entered in lower case, ensure all the data in that field is lower case.

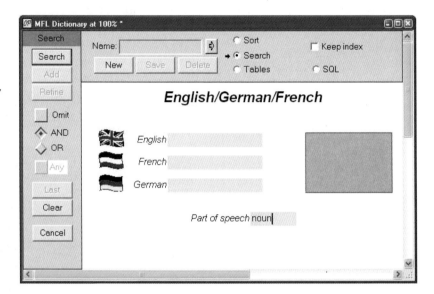

Once again, the exact method of initiating a search will vary from program to program. In the example shown here, choosing Search from the menu opens a window showing a blank record card. If you want to find the translation for 'run', enter the word into the 'English' field and click the Search button. If you have a German word but don't know what it means, enter the word into the German field.

Some database searches are 'case-sensitive'. In other words they would treat Noun and noun as two different words.

Entering a word in the Part of Speech field (in this case, noun) will list all of the words in your database which you have declared as being a noun.

Mathematics

One of the key computer programs for mathematics is the spreadsheet which is arguably the most powerful of the Office Suite of applications.

Covers

Chapter Seventeen

Spreadsheets

You will probably be given many problems that you can solve using a spreadsheet. Here is a simple shopping list showing a number of items with their prices and the quantities.

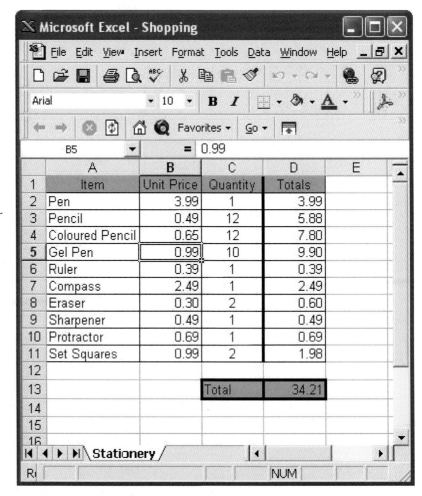

The symbol for multiply is an asterisk which you can enter using Shift+8.

To explain how it works, it's probably best to recreate it yourself.

1 Start your spreadsheet program and open a new document.

2 Copy columns A, B and C exactly as they are shown above.

You'll probably have to increase the width of column A in order to fit in the items.

3 If you wish you can colour the cells in row I and add some lines. (Right-click on row I and choose Format.)

Now comes the interesting part. The numbers in column D are actually the results of calculations performed by the computer.

Don't believe everything a computer tells you. Check it – to ensure the computer has got it right.

4 Click on cell D2 and enter the formula =B2*C2 and press Return.

You will see that the result of that formula (the contents of cell B2 multiplied by the contents of cell C2) appears in cell D2.

5 Now click on cell D3 and enter the formula =B3*C3 and press Return.

Once again the result of the formula (the contents of Cell B3 multiplied by the contents of cell C3) appears in cell D3.

You could continue typing in the formula for each of the cells in column D, but there is a quicker way using the Copy and Paste facility.

6 Click on cell D3 and press Ctrl+C to copy the contents into the temporary memory area called the Clipboard.

Before copying and pasting, try to predict what each formula will be.

7 Now click on cell D4 and drag down to DII.

8 Press Ctrl+V and the formulas will be pasted into each cell.

The really clever part is that the formula has not been copied exactly. The original formula that was copied was =B3*C3. If you look at cell D8, for example, you'll see that the formula has been altered to =B8*C8.

The total shopping bill is shown as £34.21p and appears in D13. The contents of this cell, too, are a formula. Before turning the page, what do you think the formula is?

You could have entered D2 + D3 + D4 + D5 + D6 + D7 + D8 + D9 + D10 + D11 + D12 which would have given exactly the same result. But using SUM is a much easier way of doing it.

Always try to find a way of using a formula rather than a number.

The formula in cell D13 is =SUM(D2:D12).

The exact method of entering the formula and the exact construction of it may vary between different spreadsheet programs, but if you type in this formula it should work for most spreadsheets.

Having completed the spreadsheet and saved it, you can now ask some *what if* questions. For example, what if the cost of Gel Pens was reduced to 75p? What if I bought an extra pen? What if pencils went up in price by 6p each and I needed 15 of them?

To answer any of these questions, simply change the appropriate number in column B or C. Each time you change a number and press Return the total on the same row and the grand total will change.

You can set other problems like, how many pencils can I buy if I also buy a pen and I only have £5.00?

Once you have created a spreadsheet you can ask all manner of questions and it will give you the answer providing you have included as many formulae in the table as you can. Any constant relationship between two cells should be able to have a formula applied to it. For example, if coloured pencils were always 1.5 times the cost of ordinary pencils, then cell B4 could carry the formula =B3*1.5. If ordinary pencils go up in price, coloured pencils will automatically go up proportionally.

Taking it further

Consider this scenario:

You have just written a book which is about to be published. Your publisher has offered you either £5000 plus 10% of the sales revenue (known as a royalty) or 12.5% royalty, or an outright one-off payment of £20,000. The book is expected to sell at £6.99 with an estimated 25,000 sales. Which is the best deal?

This is always a difficult choice for a writer because it's not possible to predict the actual sales, but assuming the publisher is correct in the estimate of 25,000 copies sold, will a 12.5% royalty on £6.99 be a good deal?

After opening your spreadsheet program and creating a new document:

Always enter the constants first.

1 | Begin entering the constants – the bits that you know about. That's £6.99 for the book, 25,000 sales, 10% royalty, 12.5% royalty and a £5000 bonus.

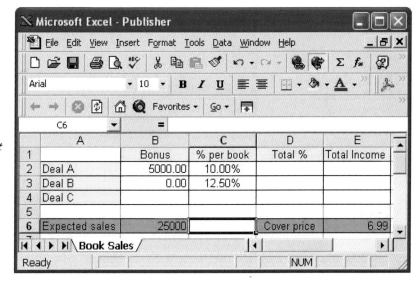

Change the number format of columns D and E so they have two decimal places.

2 | The total % income for Deal A will be the price of the book (E6) multiplied by the % royalty per book (C2) multiplied by the expected sales (B6). So in cell D2 you should enter =B6*C2*E6.

Don't include symbols like £.

3 | The total income for Deal A will be the Total % (D2) plus the Bonus (B2). So in cell E2 should go =B2+D2.

4 | Deal B should then be worked out in the same way as Deal A.

5 | The flat rate of £20,000 doesn't require any calculations because that figure is constant. Enter 20000 in E4.

When submitting your final answer there are other factors that you might want to take into account.

For example, speaking mathematically, how much interest would you receive on £20,000 if invested in a Building Society?

In a non-mathematical sense, a one-off payment of £20,000 might be of more benefit than several smaller amounts coming in over a period of time...

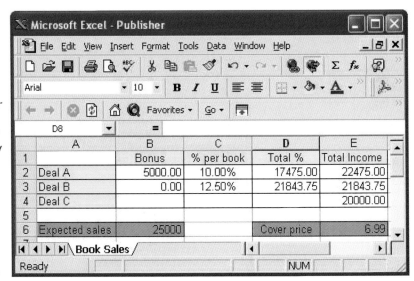

As you can see from the spreadsheet, based on the figures provided by the publisher, Deal A is the best choice.

But what if your book doesn't do as well as expected. Suppose you only sell 20,000 copies. Which would be the best deal then? Or perhaps you sell 30,000 copies, would Deal A still be your first choice?

Having established which deal is best for those figures, your publisher contacts you and says that he thinks a cover price of £5.99 instead of £6.99 would sell more copies. How many more copies would you have to sell to get the same financial return?

Your expenses

For further practice, you could try creating a spreadsheet of your weekly or monthly finances. You could enter your weekly income, and list your expenses under headings such as travel, school equipment, clothes and entertainment.

You could then calculate how long it would take you to save for something from the money you have left over each week.

Function graphs

As well as displaying results in a table, spreadsheets can plot graphs based on mathematical functions.

The ^ symbol is obtained by pressing Shift +6. It means 'raise to the power of'. Immediately after you must enter a number – in this case it is 2, which means raise to the power of 2 which is usually referred to as 'squared'.

1 In cell A2, enter the number -5. In A3 write the formula =A2+1 and copy it down to A12. This should give you numbers in the range -5 to +5.

2 In cell B2, enter the formula =A2 ^ 2 and copy this down to B12.

We start on row 2 so that the titles can be entered in row 1.

3 Select the two columns, click the Graph creation button and choose a line graph to plot a graph of the function.

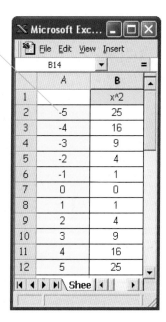

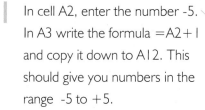

To copy a formula, select the cell with the formula, press Ctrl+C. Click and drag the mouse pointer over the cells you want to copy the formula to and press Ctrl+V.

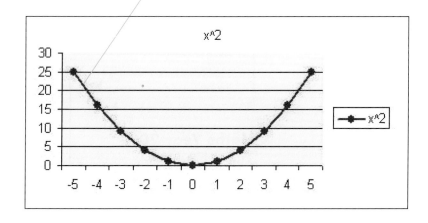

Try entering =A2 ^ 2-5 into B2 and copying down to B12. Or how about something really clever like =(A2-2)*(A2+3)*(A2-7)?

Websites

It's worth looking on the Web for Maths help. There are several sites which feature Maths games and puzzles to solve.

NRICH

This site provides additional help for most levels. If you get stuck you can email a mathematician to get the answer. There's also a good selection of puzzles to get you thinking. The NRICH website is at http://nrich.maths.org.uk/.

Put these sites into your Favorites folder.

Maths Year 2000

It's easy to spend too long on these sites and forget your homework.

This website at www.mathsyear2000.org/ is another well thought-out site with lots of fun activities for all levels. Follow the links to the Explorer and you'll find some really interesting articles about a wide range of mathematical topics including fractals, morphing and random numbers. There's also the popular Daily Puzzle and some bits of general interest like mathematical history.

You can also access this site at www.counton.org/.

Maths Is Fun

This superb site at www.mathsisfun.com/ has a wide range of problems, games, puzzles, and offline activities, together with a discussion board and a monthly newsletter.

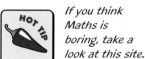

If you think Maths is boring, take a look at this site.

If you have a problem with almost any aspect of your Maths work, visit this site and follow the signs to the Index. Click on the initial letter of the Maths topic and then choose the topic from the list. You'll find yourself on a page with clear instructions, diagrams and some problems to solve.

Function plotting

There are several sites that offer additional help with plotting functions. One of the most popular sites is at http://mss. math.vanderbilt.edu/~pscrooke/toolkit.shtml. When you get into the site, choose Graphs of Functions from the menu.

No homework?

Tessellations

The Dutch artist M C Escher studied tessellations and produced many famous pictures based on the principle of interlocking shapes. Your studies of two-dimensional shapes will probably include tessellations. There are computer applications available which are specifically designed for creating tessellations but if you have a vector drawing program, you can experiment with simple tessellating shapes.

The simplest tessellating shape is a square and that's probably the best starting point.

1   Open a new document and set a fairly large grid – say .5mm with no subdivisions.

2   Draw a square, but not using just four lines. As you draw each side of the square, click on every point on the grid.

The total area of the final shape will be the same as the square on which it is based.

Once you have produced one shape, make copies of it and drag them into position to see if it really works.

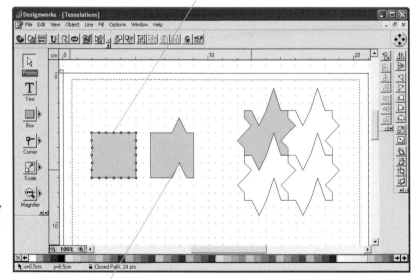

3   You can now drag individual points, but for every point you move, you must move the opposite point in the same direction and by the same amount.

Number series

The numbers 2 4 6 8 are called an arithmetic series. Each pair of numbers is separated by a common rule. In this example, the rule is Add 2. You can use a spreadsheet to prove this:

1   Create a new sheet and in A1 put the number 2. That should be the last number you enter – the rest will be formulae.

2   In cell A2, put the formula =A1+2.

3   Now copy and paste the formula into the other cells in column A.

Column A should now have the numbers 2 4 6 8 10 12 etc.

Use a spreadsheet to find the rules for the following and find what the next two numbers in each series will be:

1 2 4 8 16 32 64 128

3 6 9 12 15 18 21 24

1 4 9 16 25 36 49 64

Now try creating a line graph for each of the series.

Try working the other way around – use a spreadsheet to create an arithmetic series.

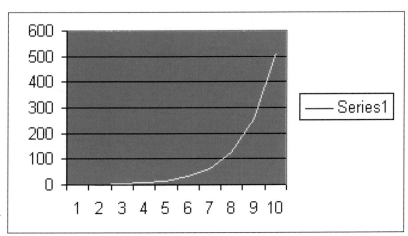

Nets

If you take a hollow 3D object like a cube and flatten it out, you are left with the net. If you want to create a 3D object from card, you must first plan the net.

There are programs specifically designed to help create nets, but you can get some really good results using a vector drawing program and some thought. To create a cube:

1   Start the drawing program and create a new document with a locked grid.

2   Draw a square, then select and copy 5 times so you have a total of six squares.

3   Drag the squares around until you have then in the form of a cross.

4   Save the drawing and then print it.

5   Carefully cut it out, score the fold lines and glue it together.

When you print your net, either use thicker paper in the printer or glue the printout onto another sheet.

If you use scissors to cut out your net, you will need to be very accurate. A better method is to use a sharp knife, a steel rule and a cutting board.

When gluing the net together, apply a thin film of paper glue along the edges. Gluing tabs is not always successful.

When you've had some practice, you can progress onto something a little more advanced like this Dodecahedron (twelve pentagons). To make one of these you'll need to choose the Polygon tool and set it to a 5-sided shape. Draw one pentagon, copy it and rotate it 180°. You'll need a total of 7 original and 5 rotated pentagons. Cut the net out really carefully.

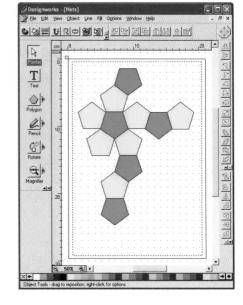

Music

If you're learning to play a musical instrument, your teacher will probably tell you that the most beneficial homework is practice, but there are other aspects of the subject which should not be overlooked.

Covers

Chapter Eighteen

Websites

Learning to read music is an integral part of learning to play an instrument.

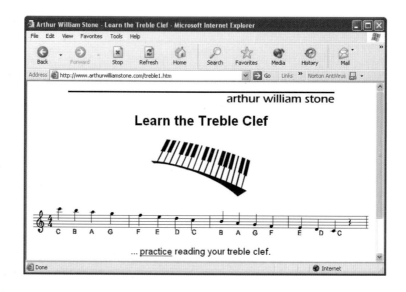

If you want to be good enough to play in a band or orchestra, you're going to need to have an understanding of music. Your music teacher may give you music theory homework and if so, and if you come across a problem with it, visit one of the many music sites like www.arthurwilliamstone.com/ for some help.

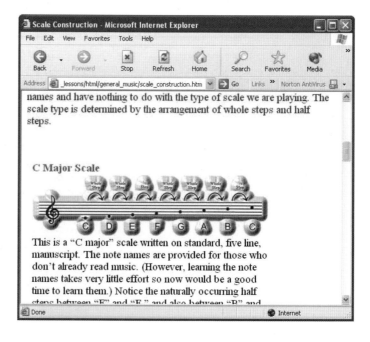

170 | Doing Homework on your Computer in easy steps

Another good site for help with music theory is Play Music Free at `www.playmusicfree.com/`. Both sites offer clear descriptions with good diagrams to help you overcome any problems you may have with music theory, and even give you a few lessons.

You can find information on almost any aspect of music on the Internet. Simply typing in the name of a musical instrument into the browser's address bar will list several. To get a list of some of the very best sites, visit `www.knowledgehound.com/`.

 There are many sites which provide specialist help for a particular instrument. Some even give away free software and/or sheet music in the form of a word processed document.

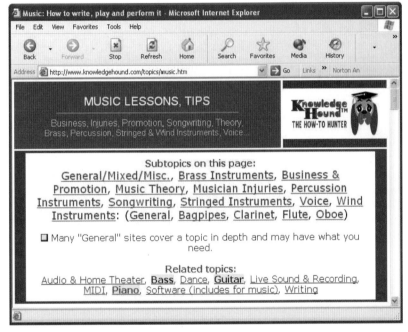

 However much help you get, it ultimately requires practice to perfect the playing of a musical instrument.

Apart from learning about instruments, learning about composers and performers is also part of your musical studies.

Most pop musicians have their own websites which provide information about the performers and the performances. The easiest way to find a particular pop site is to visit the VH1 site at `http://artists.vh1.com/` where you can browse the index of artists listed under the type of music they play. Alternatively you can enter their name into the search panel.

No homework?

Sound recording

Recording studios no longer use tape, but save digitally recorded sounds onto a computer. There are several recording programs available for your PC, but if you want to experiment without spending a great deal of money, you can get an insight into the basics with Sound Recorder which is supplied with Windows.

Keep the microphone away from the speakers or you'll get feedback: a high pitched scream which is the result of the microphone picking up the sound from the speaker.

In order to use this software you will need a microphone and a reasonably good pair of speakers. Most computers are now supplied with speakers but if you haven't got a microphone you can buy a cheap model from most high street computer/hi-fi stores. But remember, you won't get high quality sound from a cheap microphone and speakers. If you try this and think you might like to do something more advanced, you might need to consider buying better sound equipment, including a new sound card that is fitted inside the computer.

It's worth spending some time getting the microphone the correct distance away from the source of the sound you are recording.

Before coming to the actual sound recorder, you need to take control of the volume and recording levels.

1. Right-click on the Volume control icon on the right of the Toolbar and choose Open Volume Control.

2. The window displays the controls for every possible input and output sound device; even ones you may not have. Begin by setting them about midway and make small changes to get the sound as clear as possible.

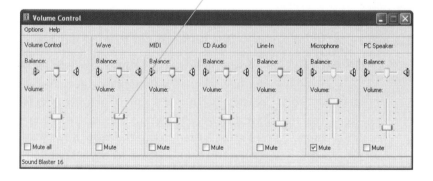

The Sound Recorder program is quite basic. Simply click the record button, sing or play into the microphone and click the stop button when you've finished.

Several music magazines come with a cover-mounted CD containing backing tracks to which you can add your own vocals or instruments.

You can then perform a limited amount of editing on the file. You may wish to remove some of the silence at the beginning and end of the file.

1 Play the recording and note where the actual music begins by referring to the displayed time on the left of the window.

2 Click the Stop button and move the slider to the position just before the time when the music is about to begin. (Don't remove all of the lead-in, leave about half a second.)

3 Open the Edit menu and choose Delete Before Current Cursor Position.

4 Do the same at the end of the recording, but this time choose Delete After Current Cursor Position.

When you have cropped the two ends of your recording, save the file.

Mixing

The previous task introduced the principle of recording sounds and putting them on your computer in the way modern recording is done. When you have created two files, you can mix them together. There is a wide range of mixing software available for your PC, but if you want to experiment without spending a great deal of money, you can with Sound Recorder:

1 Open the first file by going to the File menu and choosing Open...

2 Move the slider to the position where you want the mixing to begin. Again, check the time on the left of the window.

3 Go to the Edit menu and choose Mix with File...

Copy	Ctrl+C
Paste Insert	Ctrl+V
Paste Mix	
Insert File...	
Mix with File...	
Delete Before Current Position	
Delete After Current Position	
Audio Properties	

4 Choose the file you wish to mix with the first.

You should now save your mixed file under a new name. Do not click on Save in the File menu as you will overwrite the first file you opened. You may need the unmixed files later.

Having saved your newly mixed file you can then mix other recordings with it in the same way. The trick is accurate queuing and that is achieved by carefully noting the times displayed on the left of the window and accurately adjusting the slider to get the files synchronised.

More sophisticated software makes the synchronisation process much easier and more accurate. You'll also be able to mix multiple files in one go rather than mixing just two at a time. But nevertheless, Sound Recorder can produce some acceptable results providing you are patient with it.

Religious Studies

This subject provides the opportunity to do some very interesting research.

Covers

Chapter Nineteen

Booklet

If you are completing Religious Studies homework on a computer, you'll probably be using a word processor or perhaps a DTP application, but rather than printing out your work on a flat piece of paper, why not present it as a booklet? It's quite easy to make an A5 booklet: it's simply a piece of A4 paper folded in half.

Because each piece of A4 paper will contain 4 pages (2 on each side) the total number of pages in your document must be divisible by 4. So, if your writing extends into page 5, you'll end up with an 8 page booklet.

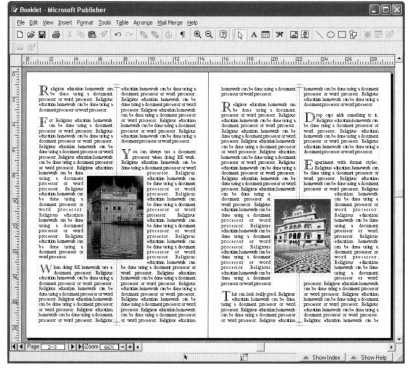

For a really good looking booklet, use two columns and keep the text size down to 10 pt.

I Use the Page Setup menu and choose a page size of 210mm high and 148.5mm wide.

2 If you have the choice, set the document to begin on an odd page.

Many DTP programs give you the option of putting a Drop Cap at the beginning of each paragraph.

The first page will be the cover and you would normally begin writing on the second page.

When you've printed your work, fold the sheet(s) in half and you'll have an A5 booklet.

Websites

Just because a religious philosophy is different to your way of thinking, it doesn't necessarily mean that they (or you) are wrong.

All religions and philosophies have several websites, some described as official and others as unofficial. Whichever category a site claims to be, they are usually crammed with information about the history of the religion and the customs. But don't take what they say on face value. Question whether you feel a particular philosophy fits in with your way of thinking.

Totally Jewish

This site at www.totallyjewish.com/ was started by a young lad in North London and is now a very popular site amongst the Jewish community. Not only does it provide information about the religion, but also the cultural side of Jewish life.

Sikh

The highlight of this site at www.sikhs.org/ is the virtual tour of the Golden Temple at Amritsar. You are presented with a plan of the temple and by clicking one of the buttons, you can download a 360° panoramic view of the area.

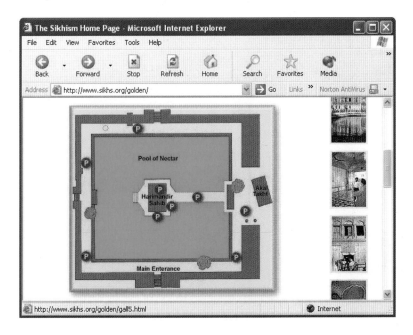

Buddhist

The website at www.lanka.net/ provides a wealth of information about this ancient Eastern religion.

No homework?

World religions

All religions have their own particular beliefs, ceremonies, philosophies and celebrations although some are common to more than one religion. It would be interesting and possibly useful to make a comparison of some of them and see where there is an overlap.

You can do this in either a database or a spreadsheet, but either way, you should begin collecting some of the information before you start.

If you want to do this in a database, each record would contain the name of the religion and the categories or fields.

1   Decide first on the fields or categories that you want to use. These could be Festivals, Life after Death, Name of Scripture, etc.

2   Next, decide on the religions you want to study. These might include Christianity, Buddhism, Judaism, Moslem, Sikh, etc.

3   If you're using a spreadsheet, put the names of the categories in row 1 beginning with B1 and the religions at the side beginning with A2.

Most spreadsheets offer the facility of locking a column (in this case column A) so that however far you move the sheet across, you can always see the titles on the left.

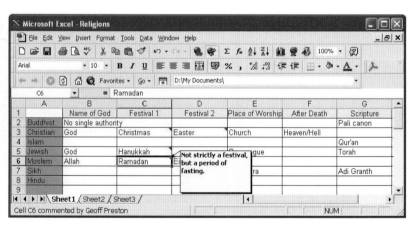

Don't be over ambitious. The simple sheet shown here has 7 religions and 6 categories. To complete this table you'll need 42 pieces of information plus any notes you may want to insert.

Once you have got the basic sheet laid out, with the titles in place, you can begin adding the information a little at a time. If your spreadsheet can have notes or comments attached to each cell, use this to add additional information about the name you've put in the cell.

Science

You'll be doing lots of experiments and investigations in science and then writing up detailed notes of your findings.

Covers

Chapter Twenty

Experiments and investigations

Your Science education will feature several investigations and experiments which will usually be followed by a detailed report. The report will normally include a diagram of the apparatus used, some calculations, possibly a table of results and sometimes a graph.

Writing up the report

Most science teachers seem to have their own preferred way of writing up an experiment so you should first check your teacher's exact requirements although they will probably be similar to this outline:

Title	Begin with a clear title at the top of the page. You could use either a larger font or even a different font style to make the heading stand out.
Task	Copy out the problem or question that your teacher has set for you. You could put this in italics.
Predictions	In this section, use your scientific knowledge and write down what results you think you will get.
Apparatus	Write down the equipment you used for the experiment or investigation.
Diagram	Include a neat, labelled diagram of the apparatus.
Method	Clearly and carefully write down how you carried out your experiment.
Results	Put your results in a table and on a bar chart or line graph.
Conclusion	In this section discuss your results.
Evaluation	Describe any problems you had and how you could improve the investigation.

Drawing the apparatus

One of the great advantages of using a computer for your work is that you can save lots of time by adapting or altering previous work.

If you have a vector drawing program, draw the apparatus used in your first Chemistry experiment. These items can then be used over and over again for different experiments. Gradually you will build up a bank of apparatus that you can use for future experiment write ups.

Keep all of your work on your computer, even after you've printed it and handed it in.

You can just about do this with a painting program, but it's not recommended.

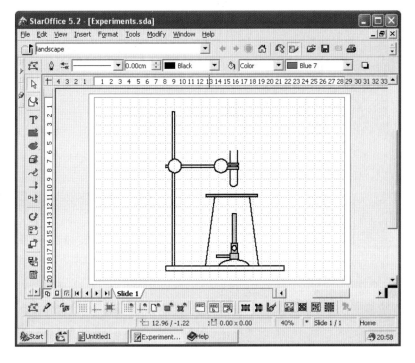

When using vector graphics, each drawing will be made from a series of components. The tripod, for example, is made from 3 components – the top is a narrow rectangle and the legs are two thick lines. When you have drawn each piece of apparatus, select all the components and group them so that they remain together.

Having created your first diagram for science, you can copy and paste parts of it into other diagrams. In the example above, having drawn a Bunsen burner, you should never have to draw it again.

Displaying chemical formulae

It's easy to display chemical formulae and equations when you know how. The chemical formula for Sulphuric Acid is H_2SO_4. You will see that the numbers are smaller and fall below the line on which the letters stand. These are called subscript characters. You can display subscripted characters in most word processors although actually creating them does vary. In StarOffice, for example:

 If your word processor can't handle subscripted characters, follow steps 1 and 2, but then reduce the print size.

1 Type the formula with capital letters and numbers e.g. H2SO4.

2 Carefully mark one of the numbers by clicking the mouse on the left of the number and dragging to the right.

3 Click on the Format menu and choose Character: to open the Character dialog.

4 Click the Sub button in the Position area and then click OK.

 If you want to type the degrees symbol (°), ensure Num Lock is on, press and hold the Alt key (not Alt Gr) and type on the numeric keypad, 0176.

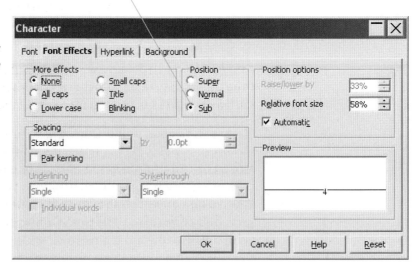

You'll now need to do the same again for the other numbers in the chemical formula.

Equations

In the context of Chemistry, an equation describes a chemical reaction and has a right-facing arrow sign in the middle. You may be asked to write the equation in either words or using chemical formulae or both.

The equation must balance – in other words you must have the same components on each side of the arrow.

$$H_2SO_4 + CuO \longrightarrow CuSO_4 + H_2O$$

Sulphuric Acid + Copper Oxide *gives* Copper Sulphate + Water

The way to get the arrow is to use a font which contains an arrow in its character set. Arial is such a font.

1. Run the Character Map program which is found by clicking Start, choosing All Programs, going to Accessories and then System Tools.

2. Choose the font from the drop-down list and select the right-facing arrow character by double-clicking on it.

3. Click the Copy button, then go to where you want the arrow to be placed and press Ctrl+V.

If you can't find an arrow in your font armoury, use a minus sign followed by a greater than sign - -> .

Another popular font for special characters is Wingdings which contains an assortment of arrows and other symbols that may be of use when writing up your science experiment.

Once you've entered the equation up to the arrow you should be able to copy bits from the left of the arrow and paste them on the right. For example, copy and paste Cu (which is part of CuO), then SO_4 (which is part of H_2SO_4) then type the + sign. Next, copy and paste H_2 before adding the final O. That will save you a great deal of time by avoiding making every individual number a subscripted character.

Table

If you have to include a table of results, use a spreadsheet. In most cases data is not entered in row 1 or column A as these are best reserved for the headings.

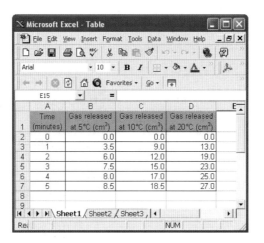

Once you've entered the data you can make the chart look more attractive by colouring some of the cells and adding borders to the cells being used.

In most cases, select the cells you wish to colour and select Format, Cells. You can then add lines, apply colour and change the format of the numbers which in most cases means the number of decimal places.

You may then be asked to produce a graph showing your results and this is something which can be done with almost no extra work. Most spreadsheets will generate a graph from the data and even include a key which it has generated. You may have to add labels for the axis if the software doesn't insert it for you.

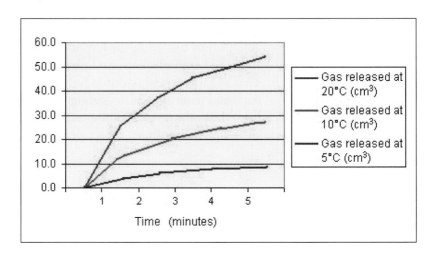

Websites

There are some great sites to help you get more information about Science topics.

BBCi – Science

The site at www.bbc.co.uk/science/ is from the Television channel and has lots of material for Physics, Chemistry and Biology.

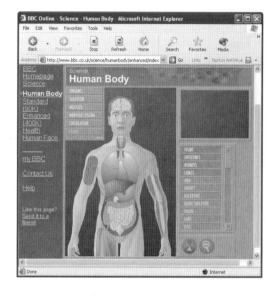

Nature News

This is the online version of the magazine that contains lots of information about nature-related issues.

There isn't a specific education site but entering a keyword in the search engine may lead to an article containing the information you're looking for. Visit www.nature.com/.

Entering 'Physics', 'Chemistry', 'Biology', or just 'Science' into a search engine will provide you with a good assortment of websites.

You could also try entering the exact topic you're studying e.g. 'photosynthesis'.

New Scientist

Another online version of a monthly magazine is at www.newscientist.com/, this site is really aimed at a higher stage of education, but there are some really interesting articles and links to other science sites.

Science Museum

The official site of London's Science Museum at www.sciencemuseum.org.uk/ is as fascinating as the museum itself. Find the link to the Learn & Teach area and then choose the Students link. You can then enter the School Stuff area.

Yuckiest Website

If you're trying to get information about human Biology, this fun site at www.yucky.com/ will provide lots of useful information.

No homework?

The Periodic Table

One of the topics you'll be studying is the Periodic Table of Elements. To help you with this work, and with your work in Chemistry in general, you might like to draw your own table and add the information about the elements as you learn it.

You could use a database for this project.

There are several ways to create a diagram of the Periodic Table but I'm going to use a spreadsheet. Although a spreadsheet is designed for calculating numbers using formulae, the fact that it's based on boxes means that you can quickly draw the table. Before beginning, check that your spreadsheet can include comments in the cells.

1. When you've opened the spreadsheet, alter the column widths and row heights so the cells are square. (Leave the top two rows as they are as these will be for the labels.)

2. In cell A1 enter the word Group and from B1 to S1 enter the numbers 1 to 18. In cell A2 put the word Period and in cells A3 to A9 enter the numbers 1 to 7:

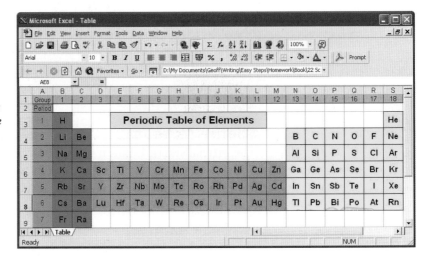

This is not the complete table showing all elements.

3. Using the picture above as a guide, add lines to the cells and then colour the cells to show the different groups of elements.

The dialog for altering cell characteristics can usually be found in the Format menu.

4 Enter the symbols for each element in the correct place.

5 Click the mouse button on one of the cells to select it, open the Insert menu and choose Comment.

6 A small box opens into which you can add further details about the element. Details could include name, atomic number, atomic weight, the name of the group the element is part of, when it was discovered, by whom, etc.

Don't try to insert all the data in one go as you'll get bored with it. Add to it a little at a time.

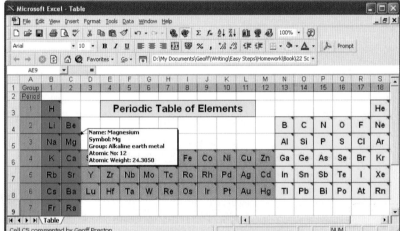

When you have entered a comment, a flag is displayed in the corner of the cell. Moving the mouse pointer over a cell with a comment will open it.

If your spreadsheet can support hyperlinks, you could link each element to a diagram showing the electron configuration.

7 If you want to draw diagrams of the elements, you can use Paint, although a vector drawing program like GST Draw would give better results.

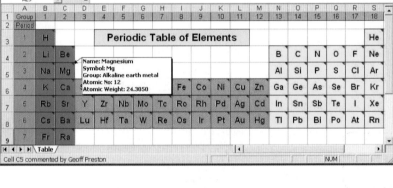

Lithium

Mixing colours

You can use a painting or drawing program to experiment with mixing colours.

1 Draw a large circle. Most applications require you to select the ellipse tool and hold down the Ctrl key to draw a perfect circle.

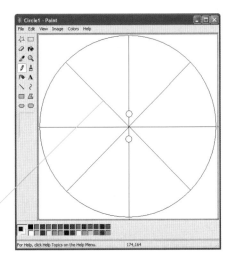

2 Divide the circle into either eight or twelve parts by drawing straight lines through the centre.

You won't get perfect mixes because of the imperfections of the printing process.

3 Colour alternate segments by choosing the colour and then choosing the Fill tool.

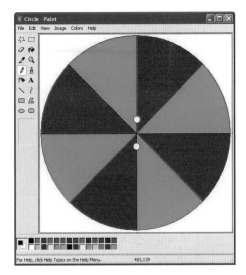

When you've completed a few, print them, cut them out and glue them onto card. Thread a loop of string through the two holes and hold the ends tightly. Now, twist the disc to make it spin back and forth quite fast which will have the effect of mixing the colours.

See what you get when you mix Red and Green, Red and Blue, Blue and Green. Now, divide the circle into 12 and try adding equal quantities of Red, Green and Blue.

Cross Curricular

All of the computer programs featured in this book can find a use in many subjects. Here are the pages that each program appears on: